studies in physical and theoretical chemistry 1

ASSOCIATION THEORY
The Phases of Matter and Their Transformations

Robert Ginell

*The City University of New York
Brooklyn College, Brooklyn, N.Y., U.S.A.*

ELSEVIER SCIENTIFIC PUBLISHING COMPANY
Amsterdam — Oxford — New York   1979

ELSEVIER SCIENTIFIC PUBLISHING COMPANY
335 Jan van Galenstraat
P.O. Box 211, 1000 AE Amsterdam, The Netherlands

Distributors for the United States and Canada:

ELSEVIER/NORTH-HOLLAND INC.
52, Vanderbilt Avenue
New York, N.Y. 10017

Library of Congress Cataloging in Publication Data

Ginell, Robert, 1912-
  Association theory.

  (Studies in physical and theoretical chemistry ; 1)
  Includes bibliographies and indexes.
  1. Phase rule and equilibrium. I. Title.
II. Series.
QD503.G55   541'.363   78-10758
ISBN 0-444-41753-2

ISBN 0-444-41753-2 (Vol. 1)
ISBN 0-444-41699-4 (Series)

© Elsevier Scientific Publishing Company, 1979
All rights reserved. No part of this publication may be reproduced, stored in a retrieval system or transmitted in any form or by any means, electronic, mechanical, photocopying, recording or otherwise, without the prior written permission of the publisher, Elsevier Scientific Publishing Company, P.O. Box 330, 1000 AH Amsterdam, The Netherlands

Printed in The Netherlands

studies in physical and theoretical chemistry 1

ASSOCIATION THEORY
The Phases of Matter and Their Transformations

studies in physical and theoretical chemistry

Other titles in this series

1 R. Ginell, **Association Theory**
2 T. Boublik, I. Nezbeda and K. Hlavaty, **Statistical Thermodynamics of Simple Liquids and Their Mixtures**
3 P. Hobza and R. Zahradnik, **Weak Intermolecular Interactions in Chemistry and Biology**
4 S. Fraga, K.M.S. Saxena and M. Torres, **Biomolecular Information Theory**

PREFACE

Association theory is not a new subject. Its genesis lies with the founders of Physics in the 1800's. However, after its auspicious beginnings its mathematical difficulties and the fact that apparently more fruitful approaches appeared led to its gradual neglect and abandonment. A few papers were published on the subject in the early 1900's and after some mention in textbooks, around 1925 it even disappeared from these works, becoming one of the discarded attempts at the explanation of physical phenomena.

I was led to this topic quite indirectly by the way of polymer kinetics, which is concerned with the conversion of the monomer to the polymer and hence the forward process. I was working at the time on the stability of some organic compounds, and the peculiar experiments I devised led me to the independent rediscovery of the higher-order Tyndall effect. The explanation of this phenomenon in turn led to a consideration of the nature of the process determining the size of particles and the equilibrium state. From this beginning, the theory evolved into an explanation of the nature of gases, liquids and solids and an understanding of the phase transformations. The mathematics is not too difficult. In some places it is very involved, but a systematic approach to the complexities of innumerable constants combined with a systematic notation has helped tame this problem. The equations follow the reasoning, and the incomprehensibility coefficient of the equations has been kept low. The book should be read systematically following the order of the chapters, since the treatment follows that order, each chapter depending on the previous ones. Some of the earlier chapters repeat in a way some of the basic material of physical chemistry and physics, but this is unavoidable since in many cases the emphasis is entirely different. It is important for even an expert in this field to read through this material, however rapidly. Later chapters are quite new and have not been treated elsewhere. The book is suitable for use as a textbook for an advanced course and as a source book for researchers in many fields, for the subjects that it covers are quite broad. Many scientists and engineers in various fields will find here topics that illuminate areas of their specialties.

Although, in general, theoretical research is a lonely field, joint thinking being well nigh impossible, nonetheless, I have had collaborators in my work. Among the many students who have worked with me, only a few were interested in theoretical work and this subject. The earliest of them was Joel Shurgan. Others were William Stein, Roseann Piser, Abraham M. Rennert, and Albert S. Kirsch. Other collaborators were Dr. Thomas A. Quigley, who did his M.A. work under my direction and later spent a postdoctoral year with me. Also Dr. Sherman D. Brown, who collaborated with me on a paper on glass and through the years since has spurred my thoughts with excellent questions and comments. My most persistent and valuable collaborator has been my wife, A. Margot Ginell, née Mayer. She was a scientist in her own right when we married and then she neglected her interest in biochemistry to concentrate on this field. She has constantly acted as a critic and sounding board for my theories and has always offered sound comments. We have published a number of joint papers. For this book she has served as editor, reading innumerable drafts and offering sound advice both on the form and content, and finally she has taken over the preparation of the camera-ready copy doing a superb job.

Also the photographs in Chapter 5 were originally published in Chemtech, the polydisciplinary monthly of the American Chemical Society ® 1972 and are reproduced with permission.

Finally, I want to acknowledge the spur to my thinking furnished by my many friends, coworkers and teachers which have directly or indirectly helped this theory grow. Among them, I particularly want to mention Professor Henry Eyring, who listened and commented adroitly, Professor Oscar K. Rice, who gave me my first job in theoretical physical chemistry and who offered pointed advice, Professor Herman Mark, who showed me the beauty of the theory of physical chemistry in his lectures, Professor

Paul E. Spoerri, my dissertation mentor, who suffered my brainstorms of undigested thoughts, and Professor Stuart A. Rice, who was my student in his undergraduate days and who worked with me for a while and whose comments I found stimulating.

Brooklyn, New York
July 1978

Robert Ginell

TABLE OF CONTENTS

	page
PREFACE	v
FOREWORD	xii
CHAPTER 1 – THE KINETIC THEORY OF MATTER	1
The Ideal Gas Law	2
Kinetic molecular theory	3
Kinetic derivation of the ideal gas law	4
Modification of the gas law	7
Van der Waals' Equation	7
The virial equation	9
Critical phenomena	9
Kamerlingh Onnes' Equation	10
A NEW VIEW OF THE KINETIC MOLECULAR HYPOTHESIS	11
The problem	11
A thought experiment	12
REFERENCES	14
CHAPTER 2 – THEORY OF ASSOCIATION	15
The problem	15
Notation	16
Formulation	17
Solution of the equations	22
THE EQUILIBRIUM CONSTANTS	28
The N notation	28
The experimental parameters	29
The conditions of constancy	31
REFERENCES	34
CHAPTER 3 – EQUATION OF STATE	35
The kinetic derivation	35
COMMENTARY ON THE DERIVATION	38
Statistical nature of collision	38
The collision distance	39
The covolume	41
The area of the wall	42
The kinetic energy of the species	42
EXPANSION TO THE VIRIAL EQUATION OF STATE	44
Elements of inversion of series	44
The f and g coefficients	45
The h and m coefficients	46

	page
The r coefficients	47
Comparison with Kamerlingh Onnes	48
The significance	49
The form of the virial coefficients	49
REFERENCES	50

CHAPTER 4 – THE NATURE OF THE ASSOCIATED SPECIES — 51

The forces between atoms	51

SYSTEMATIC DESCRIPTION OF THE j-MER — 53

The representation of a bond and the 2-mer	53
Linear forms and the 3-mer	54
The 4-mer	54
The 5-mer and model building	56
The 6-mer and transition between forms	56
The 7-mer: a and t forms	59
The 8-mer and higher forms	61
Model building of the higher forms	61
The 19a-mer and 19t-mer	62
Symmetry: Exact and inexact, 5-and 6-	63
Bernal's work	65
REFERENCES	66

CHAPTER 5 – PHASE CHANGE — 68

GROWTH AND DEGRADATION — 69

Collision of a 1-mer with a 2-mer	69
The 1-mer and the 3-mer or the 3-hole process	71
The 4-hole process	72
The 5-hole and higher processes	73

THE TRANSITION OF THE GAS TO THE LIQUID — 73

The size of m	73
At higher temperatures	75
At lower temperatures	75
A new phenomenon	76
The heat relationships	77
Mathematical consequences of the formation of a gap	77
The gas and the liquid	78
The liquid-gas phenomenon – boiling	78
The critical state	80
The value of α at the critical point	81
Evaluation of N	83
Value of N_1	84
The compressibility factor	88

THE LIQUID-TO-SOLID TRANSITION AND NUCLEATION — 88

Symmetry	89
Heterogeneous nucleation	90
Homogeneous nucleation	90

OTHER TRANSITIONS — 92

The direct gas-to-solid transition	92
Sublimation (solid-gas-solid)	92
The melting process	92

	page
Solid state transitions	93
Supersaturation in vapors	93
REFERENCES	95

CHAPTER 6 – ENTROPY · 96

The first law	96
The second law	98
Relationship to association theory	100
The transition state	101
Quantitative relationships	102
The dilemma of $\Delta\sum N_x$	104
The equation of state revisited	105
Entropy and the second law	106
Bonds and structure	106
REFERENCES	113

CHAPTER 7 – THE LIQUID STATE AND SURFACE TENSION · 114

Equation of state of the liquid	114
The surface layer	115
Surface tension	118
IMPORTANT EQUATIONS	126
REFERENCES	126

CHAPTER 8 – –THE TAIT-TAMMANN EQUATION · 128

The derivation	129
The constant L	130
The constant J	132
Determination of J and L	133
Agreement with experimental data	142
Value of P_i	143
The surface area	145
Value of the covolume B	145
Molecular volume of methanol	148
Value of the factor B	151
IMPORTANT EQUATIONS	153
REFERENCES	153

CHAPTER 9 – THE LIQUID AND THE SOLID STATE · 156

The saturated volume	156
Degree of aggregation at the saturation point	157
The maximum in B	166
The values of normal helium 4	170

THE SOLID STATE · 173

The Tait-Tammann Law and solids	173
The alkali metals	175
The internal pressure of solids	177
Atomic radii	181
Other solids	183
IMPORTANT EQUATIONS	186
REFERENCES	187

		page
CHAPTER 10 - THE CRITICAL STATE		189
The phenomenological point of view		189
CRITERIA OF THE CRITICAL STATE		195
The density and the refractive index		196
The surface tension		197
The differential criterion		198
The Tait-Tammann equation and the critical point		199
The surface tension, L and the critical region		201
The ammonia data		202
Diethyl ether		205
Water		206
Conclusions		207
REFERENCES		209
APPENDIX		211
LAGRANGIAN INTERPOLATION SUBROUTINE		211
LAGRANGIAN DERIVATIVE SUBROUTINE		213
DEMING ANALYSIS SUBROUTINE		215
The Tait-Tammann coefficients		215
THE SUBROUTINE VAR		217
CONVERGENCE		218
REFERENCES		219
AUTHOR INDEX		220
SUBJECT INDEX		222

FOREWORD

It has been said, as an aphorism, that formulating a question is half the answer. Like most aphorisms, this has an element of truth in it but in its cleverness it overstates its case. The implicit assumption here is that, since in formulating a question, half has been accomplished, the other half should be forthcoming just as readily. The half in the statement is not a mathematical half but a rhetorical one and the two halves are not equal. We all know there are many problems which, while capable of being formulated, are not solvable. Of course the aphoristic answer to this is, that we have not formulated the question correctly; thus we are not half-way there. Only when we have the answer, do we know that we have asked the right question. Looked at from this point of view, we really cannot ask the right aphoristic question before we have the answer. That is, the right question, which points directly to the answer, is generally a matter of hindsight. The fact of the matter is that we do not ask the right question ever but rather ask a multitude of relevant and irrelevant questions and stumble and stagger until one bright morning an answer appears, imperfect and tentative. After the answer has been tested and refined by further stumbling and bumbling, only then do we perceive the whole, encompassed by the answer; and only then can we formulate the ideal question demanded by the aphorism. Both the question and answer then become obvious and almost trivial and readers wonder why so much effort was needed for such a solution.

In general, there are two ways in which to present an exposition of a theory. The first is the historical one. This is an approach that reproduces the stumbling and bumbling of looking for an answer; that explores all the blind alleys along the way; that reproduces the original path of a rat going through a maze for the first time. The second method is a more didactic approach that takes us directly through the maze; and only after we have seen the final answer does it go back to explore interesting byways.

Much can be said for both methods. The historical approach gives us an insight into the mind of the founders and illuminates the creative process in the milieu of the thinking of the times. On the other hand, the didactic process saves considerable time in its exposition, is simpler and more direct. But it creates in its readers a feeling that the development is all magic, supernatural; that it was all born fully mature like Athene from the head of Zeus. However, if the aim is insight and brevity, the choice must be the didactic approach. Hence in this exposition the didactic approach will be used, with frequent references to the historical background.

CHAPTER 1

THE KINETIC THEORY OF MATTER

Before we plunge into the mathematical intricacies of this theory, a little perspective is necessary so that we shall know where we have been and hence where we are going. Like all history, this is a treacherous venture, since we have been to so many places and encountered so many ideas that the past looks different from every vantage point. The path that I will retread will not be complete, and many may quarrel with my choices; the complete story will be told one day by a more conscientious historical scholar.

The beginning of the problem of how to describe the nature of matter lies buried in the dim darkness of the past in a semantic wilderness. We will start much later in what we call the Renaissance. To be more specific, the beginnings of the solution undoubtedly start with the formulation by Boyle in 1620 of his famous law, which states that the volume of a gas is inversely proportional to the pressure, or symbolically $v \propto 1/p$.

The next pertinent step in the process occurred a considerable time later, the interval being occupied by the development of the concept of temperature in quantitative terms and the invention of a practical measuring device, the thermometer. This step was the formulation of the famous Gay-Lussac's or Charles' Law. This law states that the volume of a gas increases linearly with the increase in the temperature, if the pressure is kept constant, or symbolically

$$v_t = v_0(1 + \alpha t) ; \qquad p = \text{constant} \qquad (1:1)$$

Here v_t is the volume at a temperature, t, v_0, the volume at a temperature zero on the scale on which t is measured, and α is a constant. The temperature, t, is measured on some practical scale. When t is zero on that scale, $v_t = v_0$. The increase in v_t is then a function of α, whose magnitude depends in part on the scale used.

This relationship was a great intellectual achievement, and efforts were soon made to define the exact value of α with much greater precision. This work included attempts to extend the range of the known temperatures. Workers soon realized that the zero on the temperature scale was quite arbitrary and that temperatures below this zero were possible. The question thus arose: was there a limit to how far below zero one could go? This question can apparently be answered with the aid of Gay-Lussac's Law by extrapolation. Namely, once $\alpha t = -1$, the equation becomes $v_t = v_0(1 - 1) = 0$. From a practical and common sense point of view then, as the temperature decreases, the volume decreased till at some negative temperature, $t = -1/\alpha$, the volume of the gas became zero. Apparently, this was the limit of negative temperatures and it was named absolute zero. Hence a new temperature scale was devised, called the Absolute Temperature Scale, where T, the absolute temperature, was related to the practical temperature by the equation

$$T = t + 1/\alpha \qquad (1:2)$$

One also sees that when $t = 0$, $T_0 = 1/\alpha$, or, converting Gay-Lussac's Law into terms of the absolute temperature, we obtain

$$\frac{v_t}{v_0} = \frac{T}{T_0} \; ; \; p = \text{constant} \qquad (1:3)$$

which is the familiar form.

THE IDEAL GAS LAW

Workers also realized about this time that Boyle's Law was valid only when the temperature was constant, hence the two laws can be written as

$$v \propto 1/p \; ; \; T = \text{constant} \qquad (1:4)$$

$$v \propto T \quad ; \quad p = \text{constant} \qquad (1:5)$$

Intuitively, it can be seen that these two equations can be combined to yield

$$v \propto T/p \qquad (1:6)$$

or, equivalently,

$$pv/T = R \qquad (1:7)$$

where R is a constant and depends on the units chosen for p, v, and T; one has only to choose a convenient set of conditions for p, v, and T and R is readily evaluated.

Unfortunately, this is where our troubles begin. With more precise measurements it was soon realized that this equation described the behavior of matter only very approximately. In certain ranges the agreement was fair, while in others it was bad. Especially important were the experiments dealing with low temperatures. Here it was soon found that the substances that were gases at normal room temperatures turned first to liquids and later to solids. Evidently the extrapolation of Gay-Lussac's Law was not valid and gases did not uniformly shrink in volume till they disappeared.

KINETIC MOLECULAR THEORY

Activity on the scientific front at this time was very intense in many fields and from this activity in the field of heat arose the kinetic theory of matter. This theory postulated that pure gases consisted of a multitude of identical particles in continuous motion. The impact of these particles on the wall of the retaining vessel gave rise to the phenomena of the pressure of a gas. The temperature was a function of the velocity, or, better, the kinetic energy of the particles. The faster the particles moved the more often they struck the wall of the container, and hence the force they exerted on the wall increased. If the external pressure was kept constant, then the walls moved outward, increasing the volume till the inner pressure (or force on the wall) equaled the external pressure, which is an explanation of Gay-Lussac's Law. Boyle's Law can be similarly explained. The idea contained in this explanation can be used to derive a descriptive equation as follows.

KINETIC DERIVATION OF THE IDEAL GAS LAW

Let us assume that we have a cubic box of side l, which contains a gas consisting of identical particles in continuous motion. Let us consider a particle whose velocity is c. We shall choose a system of rectangular coordinates, x, y, z, such that one corner of the box is at the origin and the edges of the box are the axes (see Fig. 1.1). The general velocity of the particle can be broken down into components in the x, y and z directions, these being u, v and w respectively. These components are related to the general velocity by the Pythagorean relationship in three dimensions

$$c^2 = u^2 + v^2 + w^2 \qquad (1:8)$$

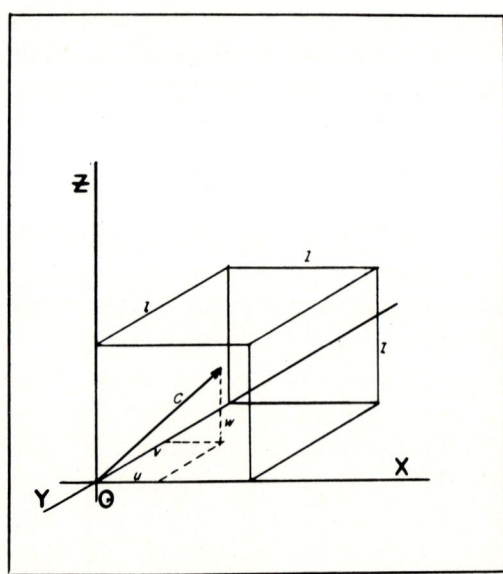

Fig.1.1. Rectangular box of side l; particle in box with velocity c has component velocities u, v and w.

The chosen particle, being in continuous motion, ultimately strikes the x wall. At the moment before it strikes it has a momentum in the x direction of mu, m being its mass. It collides with the wall and rebounds elastically, its momentum after collision being $-mu$, the velocity having changed sign, since its direction in the x coordinate has been reversed. The momentum and velocity in the y and z directions are not affected. The total change in momentum that the particle has undergone is

$$2\ mu \quad (= \text{change in momentum})$$

The particle now continues in motion till it strikes the opposite x wall. It may in the process strike also the y and z walls but since the coordinates are independent, this is immaterial to the argument at the moment. The time it takes before it strikes the opposite x wall is l/u seconds and the number of collisions it makes per second with the original wall is $u/2l$. Then the change in momentum per second at this wall is

$$\frac{u}{2l} \cdot 2mu = \frac{mu^2}{l} \tag{1:9}$$

This is not the only particle striking the wall; others also strike the wall. In fact, n particles strike the x wall per second. However, the velocity and hence the momentum of each particle is different. Thus we shall take an average, and the total change of momentum at the x wall is $n m \overline{u^2}/l$, where the average is that of u^2. This change of momentum exerts a force on the x wall, and the pressure, being defined as the force per unit area,

$$p = F/l^2 \tag{1:10}$$

yields a pressure on the x wall of

$$p_x = \frac{n m \overline{u^2}}{l^3} \tag{1:11}$$

Since the pressure on all the walls is the same, then the average of the square of the velocities in each direction x, y, z must be equal, or

$$\overline{u^2} = \overline{v^2} = \overline{w^2} \qquad (1{:}12)$$

Therefore, since

$$\overline{c^2} = \overline{u^2} + \overline{v^2} + \overline{w^2} \qquad (1{:}13)$$

$$\overline{u^2} = \frac{\overline{c^2}}{3} \qquad (1{:}14)$$

and, since l^3 equals the volume, V, of the box,

$$p = \frac{nm\overline{c^2}}{3V}$$

or

$$pV = \frac{nm\overline{c^2}}{3} \qquad (1{:}15)$$

Now the derivation becomes a little less direct. On the one hand we have the empirical relationship, $pV = RT$, on the other, the derived relationship, $pV = nm\overline{c^2}/3$. If the equations are to be consistent, it must be that

$$RT = nm\overline{c^2}/3 \qquad (1{:}16)$$

or in words, that T is a function of the average of the squares of the velocities, or, if one allows molecules of differing molecular weights, then a function of $m\overline{c^2}$. Usually the relation is stated another way. Since the kinetic energy of a single particle is given by mechanics as K.E. = $\tfrac{1}{2}mc^2$, the term $\tfrac{1}{2}nm\overline{c^2}$ is set equal to the average kinetic energy, i.e.,

$$\text{K.E.} = \tfrac{1}{2} nm\overline{c^2} \qquad (1{:}17)$$

or

$$RT = \tfrac{3}{2} \text{K.E.} \qquad (1{:}18)$$

This relationship in turn gives rise to the Law of Equipartition of Energy. The reasoning is that we have 3 directions in space and the energy must be the same in each direction. Each direction is independent, hence the kinetic energy in each direction is $\tfrac{1}{3}$ the total or

$$\text{K.E.} = \tfrac{1}{2} RT \quad \text{for one degree of freedom} \qquad (1{:}19)$$

MODIFICATION OF THE GAS LAW

We have thus used the fundamental principles of mechanics to derive the gas equation previously derived empirically from Boyle's and Gay-Lussac's Laws. However, the fact that we have given a formal derivation to the law does not change the fact that the law is obeyed only very approximately. The conclusion must be drawn that something of vital importance has been omitted from the theoretical derivation, but what was omitted was not at all obvious. However, because this law was the only one available and no gas really obeyed it, scientists postulated a hypothetical gas which did obey this law and named this the Ideal Gas or the Perfect Gas and correspondingly, the law was called the Perfect Gas Law or the Ideal Gas Law.

In a way this is an amazing piece of impudence. There exist in nature gases with certain natural properties. We devise a law to describe their behavior. The gases do not behave as predicted by the law, hence the gases are bad, imperfect; a perfect gas would obey the law; certainly, impudent anthropomorphism. Presently, real gases are named imperfect or non-ideal gases and great effort has gone into trying to correct for these imperfections. Perhaps the semantic implications inherent in this remnant from the Victorian era have impeded progress in this field.

VAN DER WAALS' EQUATION

In any case, the next efforts in this field were semi-empirical. Scientists realized that certain assumptions had been implicitly made in deriving the "Perfect" Gas Law. One of these was that, for purposes of the derivation, the molecules were considered as point-masses, i.e., that the molecules had mass but no volume. The realization that the particles of the gas must have a volume, and a rather incompressible volume, came from the fact that gases, on being compressed, turned into liquids that were well-nigh incompressible. Hence, Clausius realized that the volume referred to in Gay-Lussac's Law was the volume of space outside the molecules; therefore the gas law should be written as

$$p(V - nb) = nRT$$

where b equals the volume of the molecules and n equals the number of formula weights. This correction offered some advance. But shortly on the heels of this improvement came the achievement of Van der Waals, who postulated that there must be an attractive force between molecules. The reason underlying this deduction is that, after the particles in the gas are compressed to form a liquid, the liquid remains in existence even when the pressure is released. The conversion of the liquid to the gas requires the input of energy into the system. He reasoned that this attraction was akin to an extra external pressure; i.e., that there is an additional pressure on the gas, p', which must be added to the external pressure. The gas law then becomes

$$(p + p')(V - nb) = nRT \qquad (1:20)$$

Further reasoning, now only of historical interest, caused him to postulate that this added pressure was inversely proportional to the square of the volume. Hence, he postulated that the gas law should be

$$\left(p + \frac{n^2 a}{V^2}\right) \cdot (V - nb) = nRT \qquad (1:21)$$

This is the celebrated Van der Waals' Equation*. When Van der Waals announced this law, he had high expectations for it, and the title of his article indicated that he thought it was the equation of state for both liquids and gases. Further measurements soon showed that the law did not hold for liquids, although it agreed better with the results for gases. It did describe, however, in a qualitative fashion the transformation of gases to liquids, but the quantitative agreement was bad. Since Van der Waals' Equation was published, many other attempts have been made to describe the behavior of the "imperfect" gases, i.e., real gases. At the last accounting I have been told that there have been some 900 equations of state published; all empirical or semi-empirical. Most apply to restricted ranges for particular substances and thus are useful but not theoretically significant.

---

* Very often the equation is written without the n's, since for one formula weight of substance $n = 1$.

THE VIRIAL EQUATION

One of these empirical laws should be mentioned in this, all too scanty, summary of the events in this important field, since it has bearing on further developments; this is the equation of Kamerlingh Onnes. Kamerlingh Onnes was a wellknown experimenter at Leiden in the field of gases and liquids at low temperatures. As part of his work he needed an equation of state and obtained it by empirical means rather than relying on imperfect theory. From well developed mathematics he knew that almost any function could be expressed as a power series. Further, Van der Waals' Equation and many of the other semi-theoretical equations could be written in the form of power series of a limited number of terms; consequently, he decided to express the gas law as a power series.

CRITICAL PHENOMENA

To understand why Kamerlingh Onnes derived his equation in this form, we must go back a little. In experimenting with gases, the peculiar phenomenon of the critical temperature was discovered. Liquids were formed from gases by compression, and the effect of temperature on the necessary pressure (for liquefaction) was investigated. It was found that as the temperature increased, the pressure needed to produce a liquid increased, but, strangely, the phenomenon did not proceed smoothly.

After a certain critical temperature was reached, no amount of pressure would condense the gas into a liquid. In other words, the pressure-temperature curve ended abruptly. This is very unusual, since natural phenomena do not usually exhibit sharp boundaries. An extensive investigation of the critical point now ensued and the critical pressure, critical temperature, and critical volume of a number of gases was determined. Using the simple gas law the value of the function $RT_c/p_c V_c$ for one mole of gas was then investigated. It was found that for most of the gases

$$\frac{RT_c}{p_c V_c} \approx 3.67 \tag{1:22}$$

This was surprising on two accounts; first, the value differed markedly from the prediction of the simple gas law, which always predicted that

$$\frac{RT}{pV} = 1$$

Second, it was understandable that the value might not agree with the ideal law; but why should it be constant for so many different gases? These ideas gave rise to the concept of corresponding states. Apparently, the same phenomena occur at the critical points of the different liquids, even though the temperature of the critical points differed. Similarly, although the temperature of the boiling points of the various liquids varies considerably, if one calculates the value of T_b/T_c, one finds that for a great number of liquids the value is approximately ⅔. This leads to the concept of reduced quantities:

$$\mathbf{p} = p/p_c, \quad \mathbf{v} = V/V_c, \text{ and } \quad \mathbf{t} = T/T_c$$

In terms of these quantities, Van der Waals' Equation became simpler, the a and b disappearing and the equation reducing to

$$\left(\mathbf{p} + \frac{3}{\mathbf{v}^2}\right) \cdot (\mathbf{v} - \tfrac{1}{3}) = \tfrac{8}{3}$$

This is called the Reduced Equation of State and is the same for all gases. Calculations on this equation have shown it to be more useful than Van der Waals' Equation.

KAMERLINGH ONNES EQUATION

Returning to this equation, Kamerlingh Onnes introduced a quantity, K, for one mole of gas

$$K = \frac{RT_c}{P_c v_c}$$

and then defined a new volume

$$v_k = \frac{v}{K}$$

This quantity can be introduced into the Van der Waals' Equation, but the result is not very useful. However, expanding on the form of this equation, Kamerlingh Onnes proposed the general empirical equation

$$p v_k = t\left(1 + \frac{B}{v_k} + \frac{C}{v_k^2} + \frac{D}{v_k^4} + \frac{E}{v_k^6} + \frac{F}{v_k^8}\right) \qquad (1:23)$$

where **B**, **C**, etc. are also series of the form

$$B = \left(b_1 + \frac{b_2}{t} + \frac{b_3}{t^2} + \frac{b_4}{t^4} + \frac{b_5}{t^6}\right) \qquad (1:24)$$

thus there are 25 constants in all. Testing of this law showed that it held with good accuracy for six gases for which data were known. Apparently, the corresponding states idea is applicable and the equation of state of real gases appears to be in the form of a series of terms. But whether this series is merely a mathematical artifice* is not known.

The Kamerlingh Onnes equation is artificial in that the odd terms in **v** and **t** are omitted, and of course, it is a finite series. Nonetheless, as we shall see later, this form really has meaning and is the natural form of the equation of state.

A NEW VIEW OF THE KINETIC MOLECULAR HYPOTHESIS

THE PROBLEM

In order to be able to render meaningful this mass of theory and empiricism and derive a law that actually predicts the behavior of matter, we must reanalyze the Kinetic Molecular Theory. According to this theory, gases consist of identical particles moving in free space, undergoing numerous collisions with one another and with the walls. The natural method of analysis of such a system appears to be the methods of dynamics, utilizing velocity and momentum. Such an analysis demands the use of time-dependent quantities. But when we are considering the properties of gases, we are generally not interested in time-dependent quantities, but rather in the equilibrium conditions where time is

---

* It is well known that many diverse phenomena may be represented by a series of polynomial terms, but the terms are usually devoid of physical meaning.

irrelevant. We need to formulate a time-independent view of the kinetic molecular theory. We shall do this with the aid of a thought experiment; i.e., an experiment which, while it cannot be performed practically, enables us conceptually to think about the problem in a meaningful way.

A THOUGHT EXPERIMENT

Let us imagine that we have a wonderful motion picture camera that has such a fine resolution and such a tremendous speed that it could take pictures of the molecules in a gas as they are moving around in space. Let us use this camera to film a motion picture of a gas. Subsequently, as we run the film through our projector at normal speed, we would see the molecules moving in space, colliding with the wall and each other. This is the picture we expect from the kinetic molecular hypothesis. Let us now do an additional experiment. Let us stop the motion entirely and examine each frame individually. Each frame would be slightly different from the other. A typical frame might look like Fig. 1.2.

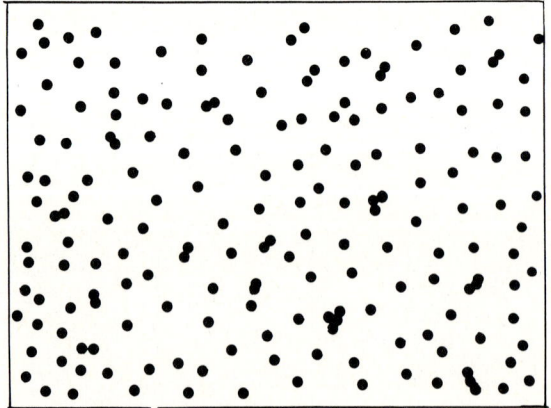

Fig.1.2. Frame of picture taken by imaginary camera which takes instantaneous pictures of substances of molecular size; 1-mers, 2-mers, 3-mers and 4-mers are shown in this picture. Part of thought experiment.

The bulk of the molecules are single, but there are some pairs of molecules in the frame; these are molecules caught in the act of collision. There are also groups of three and four, etc. These are higher order collisions. This is the conventional view. Let us now skip a large number of frames and look at another single frame. The appearance of this frame is almost entirely different; the particles are in a totally different array. Let us examine more and more frames. Each is different, yet there is a constant factor present in all the frames. Each frame can be considered a sample of the gas and we assume that many particles are in each. If we count the number of single molecules, the number of pairs, the number of threes, fours, etc. in each frame, we shall find that the number is constant. This is because the gas is at equilibrium and each volume (each frame, each sample) contains the same number of collisions of two, of three, of four, etc.

Here is our clue to eliminating the factor of time from the kinetic molecular hypothesis. We shall consider that a gas consists of an ensemble of associated molecules, i.e., of a fixed number per unit volume of single particles (1-mers), double particles (2-mers), triple particles (3-mers), etc. We shall not limit the size of these associated groups, since all sizes are possible. A particle in general terms without specifying its size will be called a j-mer. Further, we do not assign a lifetime to these j-mers; we do not know and at equilibrium do not care, since the lifetime is a time-dependent quantity. We say therefore, that gases at equilibrium consist of an array of j-mers, the number of each individual species being constant, if the temperature is constant. We now need the answers to two general questions. First, what are the relative numbers present of each of these species? Second, what is the gas law connecting them with the external observable parameters: pressure, volume, and temperature? The succeeding chapters will answer these questions.

REFERENCES

Boks, J.D.A. and Kamerlingh Onnes, Onnes Communications, No.170a, Leiden, 1924.

Boltzmann, L., *Ableit. d. Stephan'schen Gesetzen betreff. d. Abhängigk. d. Wärmestrahl v. d. Temper., aus der elektromagnet. Lichttheorie,* Ann. Physik [3], 22 (1884) 39.

Clausius, R., *Die Kinetische Theorie der Gase,* F. Vieweg und Sohn, Braunschweig, 1889, Chap.II.

Ginell, R., *Association and the general equation of state,* in *Advances in Thermophysical Properties at Extreme Temperatures and Pressures,* Amer. Soc. Mech. Eng., New York, 1965, pp. 41-48.

Glasstone, S., *Textbook of Physical Chemistry,* Van Nostrand, New York, 1940.

Jeans, J.H., *An Introduction to the Kinetic Theory of Gases,* Cambridge University Press, Cambridge, 1940.

Jeans, J.H., *The Dynamical Theory of Gases,* 4th edn. 1925, reprinted by Dover, 1954.

Natanson, L., *Kinet. Theorie d. Dissociat. Erschein. in Gasen,* Ann. Physik [3], 38 (1889) 288.

Thomson, J.J., *Chemical combination in gases,* Phil. Mag. [5], 18 (1884) 233.

Van der Waals, J.D., *De continuiteit v. d. vloeibaaren en gasvormigen toestand,* (Diss.) Leiden, 1873 (trans. into German by F. Roth, Leipzig, 1881).

CHAPTER 2

THEORY OF ASSOCIATION

Although we are primarily interested in molecules and how they associate, the theory of association is at the outset really a mathematical problem. Since the solution of this mathematical problem can have wider applications, we shall treat it here as such and only later become more specific.

THE PROBLEM

Let us assume that we have a large but finite number of objects or units of identical denomination. These units are called 1-mers. These objects can combine to form larger denominations, 2-mers, 3-mers, 4-mers, etc. Objects of higher denominations can also break down to form objects of lower denominations. In this particular case the 1-mer is the smallest object that can exist. The size of the largest object, the m-mer, i.e., the magnitude of m, depends on a number of factors which will be considered later. We thus have a dynamic process.

We start with a collection of 1-mer units or objects and change the conditions under which they exist, so that they now start to aggregate into larger j-mers, where j stands for a number greater than one. These larger j-mers in turn, once they are formed, can either continue to grow or they can degrade to smaller objects. The problem can be stated thus. If the number of unit objects, 1-mers, with which we start, is fixed and the process continues for a long time until equilibrium is established, what will be the number of objects of each denomination, i.e. of each size, that exists at that time?

It must be realized that this is a statistical problem in the sense that, if we had a limited number of objects, we could not word the question this way. As an example, if we started with only six 1-mers, then after a while we might have a 1-mer and a 5-mer. But since this is a continuing dynamic situation, at some later time the situation might change to a 2-mer and a 4-mer, or a 1-mer, a 2-mer and a 3-mer, or some other configuration of six objects. In such a case we will have to consider the probabilities of the various configurations. If we start with a large number of 1-mers, then we can talk of an equilibrium state, which would be the most probable state.

NOTATION

In order to solve this problem we shall go to a more compact mathematical notation. The reactions which yield these higher j-mers can be written as follows:

$$1\text{-mer} + 1\text{-mer} \rightarrow 2\text{-mer}$$
$$2\text{-mer} + 1\text{-mer} \rightarrow 3\text{-mer}$$
$$3\text{-mer} + 1\text{-mer} \rightarrow 4\text{-mer}$$
$$2\text{-mer} + 2\text{-mer} \rightarrow 4\text{-mer}$$

or in general

$$r\text{-mer} + s\text{-mer} \rightarrow q\text{-mer, where } q = r + s.$$

These are the growth equations. Similarly, the degradation reactions are as follows:

$$2\text{-mer} \rightarrow 1\text{-mer} + 1\text{-mer}$$
$$3\text{-mer} \rightarrow 2\text{-mer} + 1\text{-mer}$$
$$4\text{-mer} \rightarrow 2\text{-mer} + 2\text{-mer}$$
$$4\text{-mer} \rightarrow 3\text{-mer} + 1\text{-mer}$$

or in general

$$q\text{-mer} \rightarrow r\text{-mer} + s\text{-mer, where } q = r + s.$$

As can be readily seen, a tremendous number of these reactions can be written, if q is large. For the formation of a q-mer, $q/2$ equations can be written, if we assume that every process is a one-step process and q is even, i.e.:

$$1\text{-mer} + (q-1)\text{-mer} \rightarrow q\text{-mer}$$
$$2\text{-mer} + (q-2)\text{-mer} \rightarrow q\text{-mer}$$
$$3\text{-mer} + (q-3)\text{-mer} \rightarrow q\text{-mer}$$
$$\vdots$$
$$q/2\text{-mer} + q/2\text{-mer} \rightarrow q\text{-mer}$$

If q is odd, then $(q-1)/2$ equations can be written. For the degradation a similar number of equations exist. If we are to handle this set of equations conveniently, a systematic notation must be introduced.

FORMULATION

We shall introduce one simplifying assumption here and justify it later. We shall consider only those aggregations that proceed by the addition of a 1-mer, and similarly, only those degradations that proceed by the production of a 1-mer. If we now denote the number of 1-mers by n_1, 2-mers by n_2, j-mers by n_j, we have the set of equations

$$n_1 + n_1 \underset{l_{1,1}}{\overset{k_{1,1}}{\rightleftarrows}} n_2 \qquad (2:1.2)*$$

$$n_1 + n_2 \underset{l_{1,2}}{\overset{k_{1,2}}{\rightleftarrows}} n_3 \qquad (2:1.3)$$

$$n_1 + n_3 \underset{l_{1,3}}{\overset{k_{1,3}}{\rightleftarrows}} n_4 \qquad (2:1.4)$$

$$n_1 + n_4 \underset{l_{1,4}}{\overset{k_{1,4}}{\rightleftarrows}} n_5 \qquad (2:1.5)$$

$$\vdots$$

$$n_1 + n_j \underset{l_{1,j}}{\overset{k_{1,j}}{\rightleftarrows}} n_{j+1} \qquad (2:1.j+1)$$

$$\vdots$$

$$n_1 + n_{m-1} \underset{l_{1,m-1}}{\overset{k_{1,m-1}}{\rightleftarrows}} n_m \qquad (2:1.m)$$

The m-mer is the largest species that can form and must exist,

---

* The 2 in Eqn.2:1.2 refers to the size of n_2. There is no Eqn.2:1.1. This simplifies the notation.

because our starting premise was that we had a finite number of 1-mers.

Since the whole process is a dynamic one, we shall write the rate equation for the process. The rate at which the 1-mers change in number is dn_1/dt, which is decreased by the rate at which the 1-mers are used up and increased by the rate at which they are formed. Stepwise, these equations are:

Rates for Eqn. no.	Rates of disappearance due to growth	Rates of appearance due to degradation
2:1.2	$k_{1,1} n_1 n_1$	$l_{1,1} n_2$
2:1.3	$k_{1,2} n_1 n_2$	$l_{1,2} n_3$
2:1.4	$k_{1,3} n_1 n_3$	$l_{1,3} n_4$
$\vdots$		
2:1.j+1	$k_{1,j} n_1 n_j$	$l_{1,j} n_{j+1}$
$\vdots$		
2:1.m	$k_{1,m-1} n_1 n_{m-1}$	$l_{1,m-1} n_m$

The k's in these terms are the specific rate constants for the particular reaction and are related to the probability that this particular step will occur. Similarly, the l's are the specific rate constants of degradation and are similarly related to the likelihood that this reaction will occur. The subscripts of k give the species that react, their sum the size of the formed species; on the other hand, the subscripts for l give the species that are formed, their sum the size of the decomposing species. The terms above may be summarized in the equation

$$\frac{dn_1}{dt} = - \sum_{x=1}^{x=m-1} k_{1,x} n_1 n_x + \sum_{x=1}^{x=m-1} l_{1,x} n_{x+1} \qquad (2:2)$$

Let us now examine the rate of change of the 2-mers. Here the process is both more complex, and simpler. 2-mers increase in two ways: by the reaction

$$n_1 + n_1 \to n_2 \quad \text{and by} \quad n_3 \to n_2 + n_1$$

and decrease in two ways:

$$n_2 \to n_1 + n_1 \quad \text{and} \quad n_1 + n_2 \to n_3$$

This means we have 4 terms

$$\frac{dn_2}{dt} = -k_{1,2}n_1n_2 + k_{1,1}n_1n_1 - l_{1,1}n_2 + l_{1,2}n_3 \quad (2:3)$$

Similarly, we can write the equation for the 3-mer and 4-mer

$$\frac{dn_3}{dt} = -k_{1,3}n_1n_3 + k_{1,2}n_1n_2 - l_{1,2}n_3 + l_{1,3}n_4$$

$$\frac{dn_4}{dt} = -k_{1,4}n_1n_4 + k_{1,3}n_1n_3 - l_{1,3}n_4 + l_{1,4}n_5$$

But before going on and writing more equations, let us look at one other point. We must consider the fact that we have a conservation law operating, i. e., the number of 1-mers with which we started was finite. This means that

$$\sum_{x=1}^{m} x n_x = n_1 + 2n_2 + 3n_3 + \ldots + mn_m = \text{constant} \quad (2:4)$$

Hence

$$\frac{d}{dt} \sum x n_x = 0 \quad (2:5)$$

Let us test our differential equations and see whether they are consistent with this condition. Let us arbitrarily say that $m = 4$ and hence our entire set consists of 4 equations. These are

$$\frac{dn_1}{dt} = -n_1(k_{1,1}n_1 + k_{1,2}n_2 + k_{1,3}n_3) + (l_{1,1}n_2 + l_{1,2}n_3 + l_{1,3}n_4)$$

$$\frac{dn_2}{dt} = -k_{1,2}n_1n_2 + k_{1,1}n_1n_1 - l_{1,1}n_2 + l_{1,2}n_3$$

$$\frac{dn_3}{dt} = -k_{1,3}n_1n_3 + k_{1,2}n_1n_2 - l_{1,2}n_3 + l_{1,3}n_4$$

$$\frac{dn_4}{dt} = +k_{1,3}n_1n_3 - l_{1,3}n_4$$

(2:6)

The terms of Eqn. 2;5 are:

$$\frac{dn_1}{dt} = -n_1(k_{1,1}n_1 + k_{1,2}n_2 + k_{1,3}n_3) + (l_{1,1}n_2 + l_{1,2}n_3 + l_{1,3}n_4)$$

$$\frac{d2n_2}{dt} = -2k_{1,2}n_1n_2 + 2k_{1,1}n_1n_1 - 2l_{1,1}n_2 + 2l_{1,2}n_3$$

$$\frac{d3n_3}{dt} = -3k_{1,3}n_1n_3 + 3k_{1,2}n_1n_2 - 3l_{1,2}n_3 + 3l_{1,3}n_4$$

$$\frac{d4n_4}{dt} = + 4k_{1,3}n_1n_3 - 4l_{1,3}n_4$$

If we now add the last set of equations, we have

$$\frac{d}{dt}\sum_1^x xn_x = k_{1,1}n_1n_1 - l_{1,1}n_2$$

Since $\sum_1^x xn_x$ is a constant by definition, the left-hand side of the equation is identically equal to zero,

$$\frac{d}{dt}\sum_1^x xn_x \equiv 0$$

The right-hand side, however, is not identically zero; hence the equations, as we have written them, must be modified.

Let us examine these two remaining terms. To cancel the $-l_{1,1}n_2$ term we need a term $+l_{1,1}n_2$. This term appears only in the equation for dn_1/dt, where it appears once; but actually the term in this equation should be multiplied by two, because for each n_2 that degrades two n_1's appear. The problem with the other term, $k_{1,1}n_1n_1$, is a little more complex. Again, the term necessary to cancel it appears in the dn_1/dt equation and is $-k_{1,1}n_1n_1$. Several points must be considered. n_1 appears twice in this term, which means that in counting the n_1's each n_1 has been counted twice (compare with a term like $k_{1,x}n_1n_x$). Hence every time the term $k_{1,1}n_1n_1$ appears it must be divided by two, and the term is $\frac{1}{2}k_{1,1}n_1n_1$. However, this does not solve our dilemma, since the $\frac{1}{2}$ appears in the terms for dn_1/dt and dn_2/dt.

Actually, the $-\tfrac{1}{2}k_{1,1}n_1n_1$ term that appears in the dn_1/dt equation should be multiplied by two. The reasoning for this statement is as follows. Let us for the moment write the change in n_1 due to this term as $k_{1,1}n_1n_1''$ (compare with a term like $-k_{1,x}n_1n_x$). However, the prime is arbitrary. Hence there is also a term $-k_{1,1}n_1'n_1$, i.e., the change in n_1 is twice that of a term like $-k_{1,x}n_1n_x$, but in this process we have counted each n_1 twice, so that the term is really

$$-\tfrac{1}{2} \cdot 2k_{1,1}n_1n_1 = -k_{1,1}n_1n_1$$

This is the term that must appear in the dn_1/dt equation. In the dn_2/dt equation, however, the term $\tfrac{1}{2}k_{1,1}n_1n_1$ appears as such due to the preceding argument.

With this modification, our equations are correct. The complete set for the 1-mer, addition and degradation, is given in Eqn. 2:7

$$\frac{dn_1}{dt} = -\sum_{x=1}^{m-1} k_{1,x}n_1n_x + \sum_{x=1}^{m-1} l_{1,x}n_{x+1} + l_{1,1}n_2 \qquad (2:7.1)$$

$$\frac{dn_2}{dt} = -k_{1,2}n_1n_2 + \frac{k_{1,1}n_1n_1}{2} - l_{1,1}n_2 + l_{1,2}n_3 \qquad (2:7.2)$$

$$\frac{dn_3}{dt} = -k_{1,3}n_1n_3 + k_{1,2}n_1n_2 - l_{1,2}n_3 + l_{1,3}n_4 \qquad (2:7.3)$$

$$\frac{dn_4}{dt} = -k_{1,4}n_1n_4 + k_{1,3}n_1n_3 - l_{1,3}n_4 + l_{1,4}n_5 \qquad (2:7.4)$$

$$\vdots$$

$$\frac{dn_j}{dt} = -k_{1,j}n_1n_j + k_{1,j-1}n_1n_{j-1} - l_{1,j-1}n_j + l_{1,j}n_{j+1}$$
$$(2:7.j)$$

$$\vdots$$

$$\frac{dn_{m-1}}{dt} = -k_{1,m-1}n_1n_{m-1} + k_{1,m-2}n_1n_{m-2} - l_{1,m-2}n_{m-1} + l_{1,m-1}n_m$$
$$(2:7m-1)$$

$$\frac{dn_m}{dt} = + k_{1,m-1} n_1 n_{m-1} - l_{1,m-1} n_m \qquad (2:7.m)$$

For this set of equations

$$\frac{d}{dt} \sum_1^m x n_x \equiv 0$$

This can readily be verified.

SOLUTION OF THE EQUATIONS

Since the system is large, although finite, a time comes when the system is in equilibrium in the thermodynamic sense. For such a system at equilibrium the number of each species can be considered constant and hence

$$\frac{dn_x}{dt} = 0 \quad \text{for all } x\text{'s} \qquad (2:8)$$

If we add the Eqns. 2:7.2 — 2:7.m we obtain

$$\frac{d}{dt} \sum_{x=2}^m n_x = 0 = \frac{k_{1,1} n_1 n_1}{2} - l_{1,1} n_2 \qquad (2:9.2)$$

Similarly, adding Eqns. 2:7.3 — 2:7.m gives

$$\frac{d}{dt} \sum_{x=3}^m n_x = 0 = k_{1,2} n_1 n_2 - l_{1,2} n_3 \qquad (2:9.3)$$

or generally

$$\frac{d}{dt} \sum_{x=j}^m n_x = 0 = k_{1,j-1} n_1 n_{j-1} - l_{1,j-1} n_j \qquad (2:9.j)$$

and last

$$\frac{d}{dt} n_m = 0 = k_{1,m-1} n_1 n_{m-1} - l_{1,m-1} n_m \qquad (2:9.m)$$

Solving these equations we have

$$n_2 = \frac{k_{1,1}}{2l_{1,1}} n_1^2 \qquad (2:10.2)$$

$$n_3 = \frac{k_{1,2}}{l_{1,2}} n_1 n_2 \qquad (2:10.3)$$

$$n_4 = \frac{k_{1,3}}{l_{1,3}} n_1 n_3 \qquad (2:10.4)$$

$$\vdots$$

$$n_j = \frac{k_{1,j-1}}{l_{1,j-1}} n_1 n_{j-1} \qquad (2:10.j)$$

$$\vdots$$

Inserting Eqn. 2:10.2 into 2:10.3 and then the result into Eqn. 2:10.4 etc. we have

$$n_2 = \frac{k_{1,1}}{2l_{1,1}} n_1^2 \qquad (2:11.2)$$

$$n_3 = \frac{k_{1,1}}{2l_{1,1}} \cdot \frac{k_{1,2}}{l_{1,2}} n_1^3 \qquad (2:11.3)$$

$$n_4 = \frac{k_{1,1}}{2l_{1,1}} \cdot \frac{k_{1,2}}{l_{1,2}} \cdot \frac{k_{1,3}}{l_{1,3}} n_1^4 \qquad (2:11.4)$$

$$\vdots$$

$$n_j = \frac{k_{1,1}}{2l_{1,1}} \cdot \frac{k_{1,2}}{l_{1,2}} \cdot \frac{k_{1,3}}{l_{1,3}} \cdot \ldots \frac{k_{1,j-1}}{l_{1,j-1}} n_1^j \qquad (2:11.j)$$

Because we are dealing with an equilibrium process, it is appropriate that we define an equilibrium constant, K, as

$$K_{i,j} = \frac{k_{i,j}}{l_{i,j}} \qquad (2:12)$$

The general term Eqn. 2:11.j then becomes

$$n_j = \tfrac{1}{2} K_{1,1} K_{1,2} K_{1,3} \ldots K_{1,j-1} n_1^j \qquad (2:13)$$

While it is often necessary to use these constants separately, it is generally convenient to use a more compact notation, thus we define a new series of K's in more compact form

$$K_2 = \tfrac{1}{2} K_{1,1}$$
$$K_3 = \tfrac{1}{2} K_{1,1} K_{1,2}$$
$$K_4 = \tfrac{1}{2} K_{1,1} K_{1,2} K_{1,3}$$
$$K_5 = \tfrac{1}{2} K_{1,1} K_{1,2} K_{1,3} K_{1,4} \quad\quad\quad (2{:}14)$$
$$\vdots$$
$$K_j = \tfrac{1}{2} K_{1,1} K_{1,2} K_{1,3} \cdots K_{1,j-1}$$

Note that the new composite K's have a single subscript, while the old K's have a double subscript. This single subscript is the sum of the double subscripts of the last term of the continued product. Eqn. 2:13, the general term, now becomes

$$n_j = K_j n_1^j \quad\quad\quad (2{:}15)$$

The general Eqn. 2:15, which we have just derived by a kinetic process, was obtained subject to the assumption that only the 1-mer process was operating. This assumption was apparently introduced arbitrarily and its use must now be justified. In actuality, in the dynamic process that is occurring in the system, there are no restrictions and not only is the 1-mer process occurring, but also the 2-mer, the 3-mer etc. processes. The 2-mer process would be the steps whereby the addition and degradation proceeds by steps of two, the 3-mer process by steps of three, etc. Let us systematically set down the equations of the process.

1-mer process	2-mer process	3-mer process	4-mer process	5-mer process
$n_1 + n_1 \rightleftarrows n_2$				
$n_1 + n_2 \rightleftarrows n_3$				
$n_1 + n_3 \rightleftarrows n_4$	$n_2 + n_2 \rightleftarrows n_4$			
$n_1 + n_4 \rightleftarrows n_5$	$n_2 + n_3 \rightleftarrows n_5$			
$n_1 + n_5 \rightleftarrows n_6$	$n_2 + n_4 \rightleftarrows n_6$	$n_3 + n_3 \rightleftarrows n_6$		
$n_1 + n_6 \rightleftarrows n_7$	$n_2 + n_5 \rightleftarrows n_7$	$n_3 + n_4 \rightleftarrows n_7$		

$$n_1+n_7 \rightleftarrows n_8 \quad n_2+n_6 \rightleftarrows n_8 \quad n_3+n_5 \rightleftarrows n_8 \quad n_4+n_4 \rightleftarrows n_8$$
$$n_1+n_8 \rightleftarrows n_9 \quad n_2+n_7 \rightleftarrows n_9 \quad n_3+n_6 \rightleftarrows n_9 \quad n_4+n_5 \rightleftarrows n_9$$
$$n_1+n_9 \rightleftarrows n_{10} \quad n_2+n_8 \rightleftarrows n_{10} \quad n_3+n_7 \rightleftarrows n_{10} \quad n_4+n_6 \rightleftarrows n_{10} \quad n_5+n_5 \rightleftarrows n_{10}$$
$$\vdots \qquad \vdots \qquad \vdots \qquad \vdots \qquad \vdots$$

Since we are at this point interested in the equilibrium state of the system, the number of each of these j-mers is constant. Hence their ratios must be constant: i. e.,

$$\frac{n_{j+i}}{n_i n_j} = K_{i,j}$$

or setting down all the equilibrium constants we have

$$K_{1,1} = \frac{n_2}{\tfrac{1}{2}n_1 n_1}$$

$$K_{1,2} = \frac{n_3}{n_1 n_2}$$

$$K_{1,3} = \frac{n_4}{n_1 n_3} \qquad K_{2,2} = \frac{n_4}{\tfrac{1}{2}n_2 n_2}$$

$$K_{1,4} = \frac{n_5}{n_1 n_4} \qquad K_{2,3} = \frac{n_5}{n_2 n_3}$$

$$K_{1,5} = \frac{n_6}{n_1 n_5} \qquad K_{2,4} = \frac{n_6}{n_2 n_4} \qquad K_{3,3} = \frac{n_6}{\tfrac{1}{2}n_3 n_3} \qquad\qquad (2:17)$$

$$K_{1,6} = \frac{n_7}{n_1 n_6} \qquad K_{2,5} = \frac{n_7}{n_2 n_5} \qquad K_{3,4} = \frac{n_7}{n_3 n_4}$$

$$K_{1,7} = \frac{n_8}{n_1 n_7} \qquad K_{2,6} = \frac{n_8}{n_2 n_6} \qquad K_{3,5} = \frac{n_8}{n_3 n_5} \qquad K_{4,4} = \frac{n_8}{\tfrac{1}{2}n_4 n_4}$$

$$K_{1,8} = \frac{n_9}{n_1 n_8} \qquad K_{2,7} = \frac{n_9}{n_2 n_7} \qquad K_{3,6} = \frac{n_9}{n_3 n_6} \qquad K_{4,5} = \frac{n_9}{n_4 n_5}$$

$$K_{1,9} = \frac{n_{10}}{n_1 n_9} \qquad K_{2,8} = \frac{n_{10}}{n_2 n_8} \qquad K_{3,7} = \frac{n_{10}}{n_3 n_7} \qquad K_{4,6} = \frac{n_{10}}{n_4 n_6} \qquad K_{5,5} \cdots$$

$$\vdots \qquad\qquad \vdots \qquad\qquad \vdots \qquad\qquad \vdots \qquad\qquad \vdots$$

For the number of the various species we have

$n_2 = \tfrac{1}{2}K_{1,1}n_1^2$

$n_3 = K_{1,2}n_1 n_2$

$n_4 = K_{1,3}n_1 n_3 \quad n_4 = \tfrac{1}{2}K_{2,2}n_2 n_2$

$n_5 = K_{1,4}n_1 n_4 \quad n_5 = K_{2,3}n_2 n_3$

$n_6 = K_{1,5}n_1 n_5 \quad n_6 = K_{2,4}n_2 n_4 \quad n_6 = \tfrac{1}{2}K_{3,3}n_3^2$ (2:18)

$n_7 = K_{1,6}n_1 n_6 \quad n_7 = K_{2,5}n_2 n_5 \quad n_7 = K_{3,4}n_3 n_4$

$n_8 = K_{1,7}n_1 n_7 \quad n_8 = K_{2,6}n_2 n_6 \quad n_8 = K_{3,5}n_3 n_5 \quad n_8 = \tfrac{1}{2}K_{4,4}n_4^2$

$n_9 = K_{1,8}n_1 n_8 \quad n_9 = K_{2,7}n_2 n_7 \quad n_9 = K_{3,6}n_3 n_6 \quad n_9 = K_{4,5}n_4 n_5$

$n_{10} = K_{1,9}n_1 n_9 \quad n_{10} = K_{2,8}n_2 n_8 \quad n_{10} = K_{3,7}n_3 n_7 \quad n_{10} = K_{4,6}n_4 n_6 \quad n_{10} \cdots$

$\vdots \qquad\qquad \vdots \qquad\qquad \vdots \qquad\qquad \vdots \qquad\qquad \vdots$

The factor $\tfrac{1}{2}$ in the leading terms is occurring again, because we are counting the n_1's, n_2's, etc. twice. The first column, by successive substitution, as before yields again the general term

$$n_j = \tfrac{1}{2}K_{1,1}K_{1,2}K_{1,3}\cdots K_{1,j-1}n_1^j \qquad (2:19a)$$

Or $\quad n_j = K_j n_1^j \qquad (2:19b)$

using our previous definition of K's with a single subscript. Now for n_4 we have two equations, one from the first column, one from the second. Since this is an equilibrium state and n_4 is constant, we have

$$n_4 = \tfrac{1}{2}K_{1,1}K_{1,2}K_{1,3}n_1^4 = \tfrac{1}{2}K_{2,2}n_2^2$$

and since n_2 is given by $n_2 = \tfrac{1}{2}K_{1,1}n_1^2$, we have

$$\tfrac{1}{2}K_{1,1}K_{1,2}K_{1,3}n_1^4 = \tfrac{1}{2}K_{2,2}(\tfrac{1}{2}K_{1,1}n_1^2)(\tfrac{1}{2}K_{1,1}n_1^2)$$

or cancelling the appropriate terms, we have

$$\tfrac{1}{2}K_{2,2} = \frac{K_{1,2}K_{1,3}}{\tfrac{1}{2}K_{1,1}} \qquad (2:20)$$

where $K_{2,2}$ is given in terms of the $K_{1,j}$ coefficient. Using this kind of reasoning, and doing the appropriate substitutions, we have, in general

$$K_{2,j} = \frac{K_{1,j}K_{1,j+1}}{\frac{1}{2}K_{1,1}} \qquad (2:21)$$

Similarly, for the n_6 terms we have 3 equations and

$$\frac{1}{2}K_{3,3} = \frac{K_{1,3}K_{1,4}K_{1,5}}{\frac{1}{2}K_{1,1}K_{1,2}}$$

and the general term is

$$K_{3,j} = \frac{K_{1,j}K_{1,j+1}K_{1,j+2}}{\frac{1}{2}K_{1,1}K_{1,2}} \qquad (2:22)$$

Similarly, for the 4-mer process, the terms are

$$\frac{1}{2}K_{4,4} = \frac{K_{1,4}K_{1,5}K_{1,6}K_{1,7}}{\frac{1}{2}K_{1,1}K_{1,2}K_{1,3}}$$

$$\vdots$$

$$K_{4,j} = \frac{K_{1,j}K_{1,j+1}K_{1,j+2}K_{1,j+3}}{\frac{1}{2}K_{1,1}K_{1,2}K_{1,3}} \qquad (2:23)$$

This can all be written much more concisely by using the single-subscript notation for the 1-mer process coefficients. Namely,

$$K_{2,j} = \frac{K_{j+2}}{K_2 K_{j-2}}$$

$$K_{3,j} = \frac{K_{j+3}}{K_3 K_{j-3}}$$

$$K_{4,j} = \frac{K_{j+4}}{K_4 K_{j-4}} \qquad (2:24)$$

$$\vdots$$

$$K_{i,j} = \frac{K_{j+i}}{K_i K_{j-i}} \quad ; \quad j \neq i \qquad (2:25a)$$

$$\frac{1}{2}K_{j,j} = \frac{K_{j+j}}{K_j K_j} \qquad (2:25b)$$

This derivation shows that our assumption is correct. If we are concerned with the equilibrium situation, we need consider only the 1-mer process. The reasons for this are, pragmatically, that we get the same equation (Eqn.19 and Eqn.15) as an answer, if we use a kinetic derivation or an equilibrium derivation. There is a more fundamental answer; at equilibrium we have m quantities, n_j (j=1,2,3,...m). Hence we must have m restrictive parameters, so that the system will be completely determined. These are the $(m-1)$ equilibrium constants, K_2, K_3, K_4, ... K_m, and the law of conservation, $\sum j n_j$ = constant.

The equilibrium constants for the j-mer processes, where j is greater than one, are all determined and available once the constants for the 1-mer process are known. This, in general, simplifies our problem, since we now have only $m-1$ constants with which to deal. The idea behind this derivation and the use of only the 1-mer process has been formulated as a principle by Blatz in his dissertation; he states "The equilibrium equations are unique, if equilibrium between all possible species is allowed no matter what the formal mechanism by which the equations have been derived."

THE EQUILIBRIUM CONSTANTS

THE N NOTATION

Let us now examine these very necessary equilibrium constants carefully and see under what conditions they are constant. The constants that we have used thus far have the absolute units of numbers. While such units may be convenient in certain mathematical cases, they are definitely inconvenient to use, if the number of particles of various sizes is very large; hence a larger unit is needed. One can pick the size of the unit to fit the problem at hand. The one usually chosen in a chemical problem is the unit of Avogadro's number, N_0. We shall use the lower case n's to represent the absolute number, while the upper case N's are in terms of Avogadro units, so that the definition is

$$N_j = \frac{n_j}{N_0} \quad \text{or} \quad n_j = N_j N_0 \qquad (2:26)$$

To convert our equations into terms of these units, let us substitute this equation into Eqn.2:12, thus

$$N_j N_0 = \tfrac{1}{2} K_{1,1} K_{1,2} K_{1,3} \cdots K_{1,j-1} N_1^j N_0^j$$

or $\quad N_j = \tfrac{1}{2}(K_{1,1} N_0)(K_{1,2} N_0)(K_{1,3} N_0) \cdots (K_{1,j-1} N_0) N_1^j \quad (2:27)$

We now see that, if we wish to preserve the form of the original equation, we must multiply each original equilibrium constant, $K_{1,j}$, by N_0, thus defining a new equilibrium constant. To differentiate between the old equilibrium constants and the new ones (and other constants to come) we will change the notation somewhat. We shall re-label the old K's using a new double symbol

$$K_{1,j} \text{ (old)} \equiv Kn_{1,j} \text{ (new)} \quad (2:28)$$

and $\quad Kn_{1,j} N_0 = KN_{1,j} \quad (2:29)$

Eqn.2:27 now becomes

$$N_j = (\tfrac{1}{2} KN_{1,1} KN_{1,2} KN_{1,3} KN_{1,4} \cdots KN_{1,j-1}) N_1^j \quad (2:30)$$

while Eqn.2:15 becomes, analogously,

$$N_j = KN_j N_1^j \quad (2:31)$$

with $\quad KN_j = Kn_j N_0^{j-1} \quad (2:32)$

We thus see that the form of all the equations is preserved, with slightly altered meanings for the symbols.*

THE EXPERIMENTAL PARAMETERS

In any system if we know the number of objects of each denomination, the equilibrium constants can easily be calculated and no special experimental problem exists. However, generally and especially in chemical systems where the number of objects is huge, this information is not directly available. The question then becomes, what information is available about the system, by which these constants can be determined? One of the fundamental

---

* Generally, the equilibrium constants in terms of absolute numbers are not used, but rather in terms of Avogadro units; hence the N in KN is often omitted.

quantities that is always known is the extent of the system, or, put otherwise, the original number of objects of unit dimensions with which we started (thinking in terms of the kinetic derivation). If we think of the system in terms of weight, then, if we know the total weight of the system and the weight of a 1-mer, m^0, we know

$$\sum_x x n_x = \frac{w}{m^0} \qquad (2:33)$$

or, in terms of Avogadro's units,

$$\sum_x x N_x = \frac{w}{M^0} \qquad (2:34)$$

where $M^0 = m^0 N_0$, w is the total weight of the system, m^0 is the weight of one 1-mer and M^0 is the weight of an Avogadro number of 1-mers. This definition is in accordance with the usual chemical notation.

While the value of w is always certain (and is in the same units as m^0 or M^0), the value of M^0, especially in a chemical case, is often uncertain, being usually based on assumptions of varying validity. More will be said about this point later.

Another quantity that may be known is the total number of objects N_j in the system. This is

$$\sum_1^m N_x = N_1 + N_2 + N_3 + \ldots N_m \qquad (2:35)$$

This quantity may often be inferred, since certain properties of the system depend on the total number of objects, while others depend on $\sum x N_x$. This latter term can be written as

$$\frac{w}{M^0} = \sum x N_x = N_1 + 2N_2 + 3N_3 + 4N_4 + \ldots m N_m \qquad (2:36)$$

The equations may be further rewritten by using Eqn.2:31, thus

$$\sum_{x=1}^m N_x = N_1 + K N_2 N_1^2 + K N_3 N_1^3 + \ldots K N_m N_1^m \qquad (2:37)$$

or

$$\frac{w}{M^0} = \sum_{x=1}^m x N_x = N_1 + 2 K N_2 N_1^2 + 3 K N_3 N_1^3 + \ldots m K N_m N_1^m \qquad (2:38)$$

Since we shall use the function defined in Eqn.2:31 and Eqn.2:34 continually and the sum notation is cumbersome, we shall re-label them using bold characters.

$$\mathbf{N} \equiv \sum_{x=1}^{m} N_x \qquad (2:39)$$

$$\mathbf{W} \equiv \sum_{x=1}^{m} x N_x = \frac{w}{M^0} \qquad (2:40)$$

$\mathbf{N}$ and $\hat{\mathbf{W}}$ are the experimental parameters.

CONDITIONS OF CONSTANCY

Let us now go back and examine the constants, KN_j, and determine the conditions under which they are constant. Let us consider that we have a system of a given $\mathbf{N}$ and $\mathbf{W}$, and that their ratio $\mathbf{W}/\mathbf{N}$ defines a related function Z, the average degree of association.

$$Z = \frac{\mathbf{W}}{\mathbf{N}} \qquad (2:41)$$

Since dividing the total number of objects of unit denomination by the number of objects gives us the average size of an object, we have here a parameter of great interest, because many properties can be given in terms of it. While both $\mathbf{W}$ and $\mathbf{N}$ are extensive quantities depending on the extend of the system, Z is an intensive quantity independent of the extent of the system. Associated with these parameters $\mathbf{W}$ and $\mathbf{N}$ are the series of constants, KN_j, which determine the distribution.

Let us now assume that we had another system twice as large with the same Z, i. e.,

$$\mathbf{W}' = 2\mathbf{W}$$
$$\mathbf{N}' = 2\mathbf{N}$$
$$Z = \frac{\mathbf{W}}{\mathbf{N}} = \frac{\mathbf{W}'}{\mathbf{N}'}$$

The main change is the size of the system; the average size is unaffected, since Z remains the same. The distribution should also be unaffected.

Let us assume that we have W and N in a volume v; if we doubled the v, we would double W and N, but Z would be unaffected, and there is no reason to suspect that the number of N_j's per unit volume is affected. In other words, the distribution is unaffected, the only change being that

$$N_j' = 2N_j$$

Let us now examine the constants KN and KN'

$$KN_j = \frac{N_j}{N_1^j}$$

$$KN_j' = \frac{N_j'}{N_1'^j}$$

since

$$N_j' = 2N_j$$

$$N_1' = 2N_1$$

$$KN_j' = \frac{2N_j}{(2N_1)^j} = \frac{N_j}{N_1^j} \cdot \frac{1}{2^{j-1}}$$

hence

$$KN_j' = \frac{KN_j}{2^{j-1}}$$

Thus it appears that doubling the extent of the system has changed the value of the equilibrium constants. This is often an inconvenient situation, since we have not changed the character of the system; Z has remained constant. This points up the fact that, while the intensive variable Z has remained constant, doubling the capacity factors W and N has changed the KN's. Obviously, the KN's are also capacity factors. We must therefore derive new K's which are intensity factors.

Let us then define a concentration C as

$$C_j = \frac{N_j}{v} \qquad (2:42)$$

Then Eqn. 2:32 becomes

$$\sum_{x=1}^{m} vC_x = C_1 v + KN_2(C_1 v)^2 + KN_3(C_1 v)^3 + \ldots$$

or $\sum_{x=1}^{m} C_x = C_1 + (KN_2v)C_1^2 + (KN_3v^2)C_1^3 + (KN_4v^3)C_1^4 + \ldots$

and to preserve form, let us state that

$$(KN_j v^{j-1}) = Kc_j \qquad (2:43)$$

or $C = \sum_{x=1}^{m} C_x = C_1 + Kc_2 C_1^2 + Kc_3 C_1^3 + Kc_4 C_1^4 + \ldots Kc_m C_1^m \qquad (2:44)$

Eqn. 2:34 now becomes

$$\frac{W}{v} = \sum_{x=1}^{m} xC_x = C_1 + 2Kc_2 C_1^2 + 3Kc_3 C_1^3 + 4Kc_4 C_1^4 + \ldots mKc_m C_1^m \qquad (2:45)$$

Let us examine these Kc's and see whether they are intensive variables. We shall proceed as previously; first we define a system having the parameters W and $C = N/v$, with

$$Z = \frac{W}{N} = \frac{W/v}{C}$$

and second a system of twice the extent but with the same Z, i.e.

$$W' = 2W$$
$$N' = 2N$$
$$v' = 2v$$
$$Z = \frac{W'}{N'} = \frac{2W}{2N} = \frac{2W/2v}{2N/2v} = \frac{W/v}{C}$$

$$Kc_j = \frac{c_j}{c_1^j}$$

$$Kc_j' = \frac{c_j'}{c_1'^j}$$

but $c_j \equiv c_j'$
$c_1 \equiv c_1'$
hence
$Kc_j' \equiv Kc_j$

Now we see that the K_c's are intensive variables and do not depend on the extend of the system.*

REFERENCES

Blatz, P.J., Ph.D. Dissertation, Princeton University, 1944.

Blatz, P.J. and Tobolsky, A.V., *Note on the kinetics of systems manifesting simultaneous polymerization-depolymerization phenomena*, J. Phys. Chem., 49 (1945) 77.

Ginell, R., *A general theory of association*, Ann. N. Y. Acad. Sci., 60 (1955) 521-40.

---

* Because the K_c notation is rather cumbersome, in succeeding chapters we shall use K to mean K_c. Any other usage will be carefully noted.

CHAPTER 3

EQUATION OF STATE

In the previous chapter the relationships between the concentrations of the various species were derived. The relationships between these concentrations and the normally measured variables, the pressure, the volume, and the temperature must now be determined.

THE KINETIC DERIVATION

Let us imagine that we have a box containing a considerable number of particles. These particles are of various masses, m_1, m_2, m_3, in general, m_j, and are moving at various velocities, c_1, c_2, c_3, in general, c_ϕ. Since the box is three-dimensional, a typical particle has components of velocity in each direction, i.e., a velocity, u_ϕ, in the x-direction, v_ϕ in the y-direction, and w_ϕ in the z-direction. The velocity vectors for a particular particle are connected by the three-dimensional Pythagorean relationship

$$u_\phi^2 + v_\phi^2 + w_\phi^2 = c_\phi^2 \qquad (3:1)$$

Hence the momentum in the three axial directions of a typical particle, m_j, are $m_j u_\phi$, $m_j v_\phi$, and $m_j w_\phi$. When this typical particle strikes the x-wall of the box, it rebounds, reversing its direction of motion in the x-direction. (The direction of motion in the y-and z-directions is not altered.) Its momenta are now $-m_j u_\phi$, $+m_j v_\phi$, and $+m_j w_\phi$, and the change in its momentum during the collision is $2m_j u_\phi$. The number of collisions that it has with the x-walls is u_ϕ/s_x, where s_x is the distance in the x-direction that it must travel between striking one wall and then the opposite wall. The force then that this particle exerts on the x-walls is the change in momentum multiplied by the number

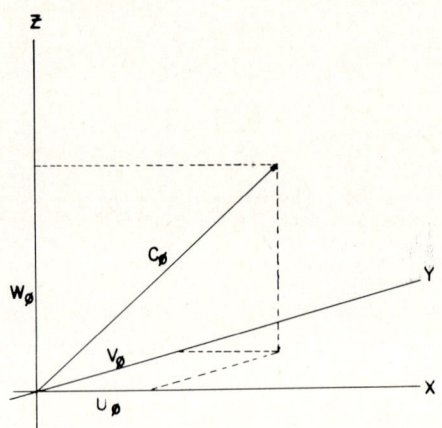

Fig.3.1. The components of the velocity vector c_ϕ.

of collisions, i.e.,

$$f_x = 2m_j u_\phi \cdot \frac{u_\phi}{s_x} = \frac{2m_j u_\phi^2}{s_x} \qquad (3:2)$$

The total force on the x-wall is

$$F_x = \sum f_x = \sum_j \sum_\phi 2 \frac{n_{j,\phi} m_j u_\phi^2}{s_x} \qquad (3:3)$$

where $n_{j,\phi}$ is the number of particles of size j and velocity ϕ. The particles of size j, since they have a velocity distribution, have an average squared-velocity of $\overline{u_j^2}$. One can replace the double sum in Eqn. 3:3 if one replaces u_ϕ^2 by the average squared-velocity $\overline{u_j^2}$ and simultaneously replaces $n_{j,\phi}$ by n_j. One then has

$$F_x = \sum_j \frac{2 n_j m_j \overline{u_j^2}}{s_x} \qquad (3:4)$$

Since the box is stationary, the forces on the y-and z-walls must be equal to the force on the x-wall or

$$\sum_j \frac{2n_j m_j \overline{u_j^2}}{s_x} = \sum_j \frac{2n_j m_j \overline{v_j^2}}{s_y} = \sum_j \frac{2n_j m_j \overline{w_j^2}}{s_z}$$

which means

$$\sum_j \frac{\overline{u_j^2}}{s_x} = \sum_j \frac{\overline{v_j^2}}{s_y} = \sum_j \frac{\overline{w_j^2}}{s_z} \qquad (3:5)$$

Since the box is symmetrical and no direction is preferential

$$s_x = s_y = s_z = s$$

and hence

$$\overline{u_j^2} = \overline{v_j^2} = \overline{w_j^2} \qquad (3:6)$$

Using Eqn. 3:1

$$\overline{u_j^2} = \frac{\overline{c_j^2}}{3}$$

since the same relationship holds for an average-squared-velocity as for a particular velocity, the force on the walls in any direction is

$$F = \sum \frac{2n_j m_j \overline{c_j^2}}{3s} \qquad (3:7)$$

Now all the particles are in equilibrium and, because of the laws of conservation of energy and conservation of momentum, the average kinetic energy of every species is equal or

$$\tfrac{1}{2} m_1 \overline{c_1^2} = \tfrac{1}{2} m_2 \overline{c_2^2} = \tfrac{1}{2} m_j \overline{c_j^2} \qquad (3:8)$$

Hence

$$F = \frac{2m_1 \overline{c_1^2}}{3s} \sum n_j \qquad (3:9)$$

The pressure on the wall is the force/area and, taking the area on which the pressure is exerted in any direction to be $2s^2$ (the 2 arising because there are 2 walls on each axis), then

$$P = \frac{F}{2s^2} = \frac{m_1 \overline{c_1^2}}{3s^3} \sum n_j \qquad (3:10)$$

Now s is defined as the distance that a particle travels in going from one wall to the opposite one. This distance is

$$s = (v - b)^{1/3} \tag{3:10}$$

where v = the actual volume of the box and b is the covolume (or excluded volume) of the particles in the box. In essence, $(v-b)$ is equal to the free volume (or space in the box). Hence

$$P = \frac{m_1 \overline{c_1^2}}{3(v-b)} \sum n_j \tag{3:12}$$

To reconcile this equation with the ideal equation we make

$$\tfrac{2}{3} \left(\frac{m_1 \overline{c_1^2}}{2} \right) = \frac{RT}{N_0} \tag{3:13}$$

where R is the usual gas constant and N_0 is Avogadro's Number. In other words, the temperature is a simple function of the average kinetic energy; thus we have

$$P(v - b) = RT \sum N_j \tag{3:14}$$

where $N_j (= n_j/N_0)$ is the number of j-mers in units of Avogadro's Number.

Although the equation resembles both the ideal equation and van der Waals' equation, there are some notable differences, which must be carefully considered. We note that b is a variable as is $\sum N_j$. The introduction of these two new variables does indeed complicate the equation, especially since they are dependent variables. But before we go on we must offer some commentary on this derivation.

COMMENTARY ON THE DERIVATION

STATISTICAL NATURE OF COLLISION

The reversal of motion in the x-direction, as usually stated, is an instantaneous reversal of the direction. The implication is that the wall has only one property, namely, that it is impenetrable and thus exerts no effect on the colliding particle other than repelling it. During this instantaneous collision, the presumption is, that the kinetic energy of the particle goes to zero instantaneously and then rebounds to its former value. If the momentum goes from $m_j u_\phi$ to $-m_j u_\phi$, it must be zero (it must stand still) for an instant, at which point the kinetic energy must be zero. These concepts are not reasonable. Considered logically, instantaneous collisions are not possible. If a reversal of momentum occurs, it must occur over a discrete interval of time; the kinetic energy must be converted to some sort of potential energy and then reconverted to kinetic energy in that discrete interval. To reconcile the concept of "instantaneous" collision with reality we must inquire more deeply into the nature of this reversal.

The impenetrable wall, since it is not imaginary, is composed of atoms which are comparable in size to the gas particles, and there is an interaction between the wall and the particles. As the particle approaches the wall, the kinetic energy of the particle is converted into a stress interaction between the particle and the wall. In the rebound, this potential energy of stress is reconverted to kinetic energy. Hence, there is a time of interaction during which the particle resides on the surface. This is a description of the process as it applies to one particle. However, since there are an immense number of particles, we can think of the process in a different way. Considered statistically at any instant, n particles are striking the wall and n particles are rebounding from the wall. This must be the case, since we are at equilibrium and the walls are not changing in thickness. For every particle that strikes the wall with a momentum of $+m_j u_\phi$, a particle with a momentum of $-m_j u_\phi$ leaves. Therefore the use of the term "instantaneous" is just a simple way of expressing a statistical theorem and as such its use is justified.

THE COLLISION DISTANCE

Another item that we must discuss, is the distance, s, that a particle travels in going from one wall to the opposite wall. If the box were to contain only one particle, this distance, s, would of course be the actual distance between walls, and s^3 would be equal to the volume, v. However, since the box contains a very large number of particles, a particle undergoes many collisions in proceeding from one wall to the other. Thus it is conceivable that a particular particle may only rarely reach the opposite wall. But looking at the derivation, we are not actually concerned with the particle traversing the box, but rather with the momentum traversing the box. The particle is just the "container" for the momentum (a momentum carrier). When a particle leaves one wall, travelling towards the opposite wall, it contains a certain amount of momentum. On collision with another particle it transfers all or part of its momentum to the other particle (see Fig.3.2). Instantaneously, this momentum is transfered through the second particle, and, because the particle occupies a discrete region of space, the distance it must travel is shortened by the distance of the diameter of this particle. ("Instantaneous" is used here in a statistical sense.) However, there are many types of particles with different diameters and the distance cannot be described so simply.

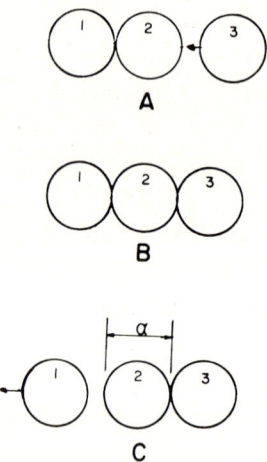

Fig.3.2. "Instantaneous" transfer of momentum.
A: before impact;   B: collision;   C: after transfer.
At any instant in time n 1-mers collide and n 1-mers leave. The lifetime of the 3-mer is not material to the argument. The transfer of momentum is "instantaneous".

THE COVOLUME

The best way of arriving at this distance is through the concept of the covolume. The covolume, sometimes called the excluded volume, is the inaccessible volume. As can be seen from Fig.3.3, the covolume of the various species is different, and since the concentration of the various species varies with the external variables, the total covolume, B, is a variable, i.e.,

$$B = b_1 N_1 + b_2 N_2 + b_3 N_3 + \ldots + b_m N_m = \sum_x b_x N_x \qquad (3:15)$$

However, b_x is a constant, being the covolume of a mole of x-mers, and

$$b_x \neq x b_1 \qquad (3:16)$$

If the total volume is v and the covolume is B, then the free volume or free space in the box is $(v-B)$, and $(v-B)^{1/3}$ is the free distance from one wall to the opposite wall.

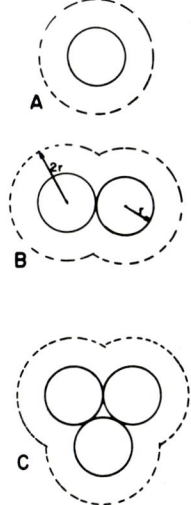

Fig.3.3. The covolume of the simpler species.
    A: The 1-mer; the dotted circle encloses the covolume.
    B: The covolume of the 2-mer.
    C: The covolume of the closely-packed (cp) 3-mer.

THE AREA OF THE WALL

In the derivation the area of the wall is taken as s^2 or $(v-b)^{2/3}$, and the distance the particle travels from one wall to the opposite wall is taken as $s = (v-b)^{1/3}$. These statements of length are in themselves not quite precise. Actually, the box, conceived for the moment as a "smooth" walled box, has an excluded volume ς; i.e., the wall excludes the particle from a certain volume (see Fig.3.4). If as in reality the wall is not smooth, the quantity ς still exists but the linear quantity β is harder to define. If we take the distance the particle travels from wall to wall as s and the area of the wall on which it impinges, as s'^2,

$$s' > s$$
$$ss'^2 \neq (v-b)$$

as given in Eqn.3:14, if we consider b to be the true covolume. Taking then the true covolume to be b' and

$$s' = s + \tau$$

then $\quad ss'^2 = s(s+\tau)^2 = (v-b')^{1/3}[(v-b')^{1/3}+\tau]^2$

or $\quad ss'^2 = (v-b') + 2\tau(v-b')^{2/3} + \tau^2(v-b')^{1/3}$

or $\quad (v-b)\{\text{from Eqn.3:14}\} = (v-b')+2\tau(v-b')^{2/3}+\tau^2(v-b')^{1/3}$ (3:17)

For a dilute gas b' is small and τ is small, so that b approximately equals b'. For a dense gas τ becomes large, $(v-b')$ becomes small, so that

$$b \approx b' \qquad (3:18)$$

Since, in general, b is calculated as an empirical factor, one must bear this inequality in mind and consider b as containing a small additional variable factor.

THE KINETIC ENERGY OF THE SPECIES

In the derivation the statement is made that "... all the particles are in equilibrium and hence, because of the laws of conservation of energy and conservation of momentum, the average kinetic energy of every species is equal". We really do not know this. What we actually mean is that all the particles have the same temperature. Unfortunately, this is a circular statement, since we define temperature in terms of the average kinetic

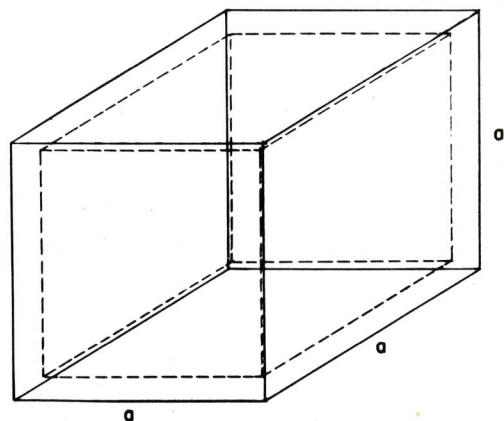

Fig.3.4. The excluded volume and the free volume of the box. The volume of the outer box is v. The distance from the walls of the inner (dotted) box to the outer wall is β. The volume between the walls of the two boxes is ζ. The relationship is

$$\zeta = a^3 - (a - \beta)^3$$
$$v = a^3$$

where ζ = excluded volume of the box.
$v - \zeta$ = free volume of the box.

energy. Systems in equilibrium are known which have particles at different temperatures. An example of such a system is a plasma, where the velocity of the electrons and ions is considerably different, and therefore the ions and electrons are considered to have different temperatures. But in this plasma system the relative sizes of the particles are greatly different, and there is a sharp gap in sizes between them. Further, the entire argument hinges on what we mean by temperature. From the classical point of view, the temperature is a function of the average kinetic energy and in general the simplest function is chosen to represent it, namely, RT, i.e.,

$$\tfrac{2}{3}\overline{KE} = RT \qquad (3:19)$$

Given the inexactitude of the temperature definition from a mechanical point of view, we feel that stating that the average kinetic energy of each species is the same, where the species are in equilibrium and closely related in size, is a plausible assumption.

EXPANSION TO THE VIRIAL EQUATION OF STATE

The derivation of a new equation of state is in itself no momentous achievement. There have been many equations of state derived for various practical and pragmatic reasons. However, these equations are generally of limited validity in the sense of being applicable over limited ranges of the external variables. We shall show that, with the introduction of these two new variables, we have derived a general equation usable over a very wide range. We shall do this in several ways. First, we shall show that this equation is the closed form of the virial equation. We know that the virial equation of state generally represents the properties of gases very well.

ELEMENTS OF INVERSION OF SERIES

The first step is to operate on

$$c = \sum_x c_x = N/v = \sum_x N_x/v \qquad (3:20)$$

and convert it into experimentally determinable functions. We shall do this by using a process called inversion or reversion of infinite series. The mathematics of this reversion is well known to mathematicians and is quite straightforward.

Assume that we have an infinite power series of the form

$$y = x + a_2 x^2 + a_3 x^3 + a_4 x^4 + \ldots \qquad (3:21)$$

The problem is to find out what x is in terms of y. In other words, if we have $y=f(x)$, what is $x=g(y)$? We shall proceed thus: Let there be another infinite series of the same form, namely,

$$x = y + b_2 y^2 + b_3 y^3 + b_4 y^4 + \ldots \qquad (3:22)$$

whose coefficients are presently unknown and are to be determined. Let us substitute the value of x from Eqn.3:22 into Eqn.3:21. Thus we obtain

$$y = (y+b_2 y^2+b_3 y^3+\ldots)+a_2(y+b_2 y^2+b_3 y^3+\ldots)^2 \\ +a_3(y+b_2 y^2+b_3 y^3+\ldots)^3 + \ldots \qquad (3:23)$$

Expanding Eqn.3:23 and collecting terms, we have

$$y = y+y^2(b_2+a_2)+y^3(b_3+2a_2b_2+a_3)$$
$$+y^4[b_4+a_2(b_2^2+2b_3)+3a_3b_2+a_4]+\ldots \qquad (3:24)$$

Equating like terms, we have

$$y = y$$
$$y^2(b_2+a_2) = 0$$
$$y^3(b_3+2a_2b_2+a_3) = 0$$
$$y^4[b_4+a_2(b_2^2+2b_3)+3a_3b_2+a_4] = 0$$
etc.

hence
$$b_2 = -a_2$$
$$b_3 = -2a_2b_2-a_3 = 2a_2^2-a_3 \qquad (3:25)$$
$$b_4 = -5a_2^3+5a_2a_3-a_4$$
etc.

Eqn.3:22 then becomes

$$x = y+(-a_2)y^2+(2a_2^2-a_3)y^3+(-5a_2^3+5a_2a_3-a_4)y^4+\ldots \qquad (3:26)$$

This type of inversion is valid for infinite series and is certainly valid for finite series.

THE f AND g COEFFICIENTS

First, we shall revert the series $\sum xC_x$*.

$$\sum xC_x = C_1 + 2K_2C_1^2 + 3K_3C_1^3 + 4K_4C_1^4 + \ldots + mK_mC_1^m \qquad (3:27)$$

obtaining

$$C_1 = (\sum xC_x)+f_2(\sum xC_x)^2+f_3(\sum xC_x)^3+f_4(\sum xC_x)^4 + \ldots \qquad (3:28)$$

where the f_x coefficients are

$$f_2 = -2K_2$$
$$f_3 = 8K_2^2-3K_3$$
$$f_4 = -40K_2^3+30K_2K_3-4K_4$$
$$f_5 = 224K_2^4-252K_2^2K_3+48K_2K_4+27K_3^2-5K_5$$

etc.

Now we shall substitute this value of C_1 into the equation for C, namely

---

* All the K's in this section are K_C's but for typographical reasons we shall use K, omitting the c.

$$C \equiv \sum C_x = C_1 + K_2 C_1^2 + K_3 C_1^3 + K_4 C_1^4 + \ldots + K_m C_1^m \qquad (3:29)$$

expanding and collecting terms we have

$$C = g_1(\sum x C_x) + g_2(\sum x C_x)^2 + g_3(\sum x C_x)^3 + \ldots \qquad (3:30)$$

where
$\quad g_1 = 1$
$\quad g_2 = -K_2$
$\quad g_3 = 4K_2^2 - 2K_3$
$\quad g_4 = -20K_2^3 + 18K_2 K_3 - 3K_4$
$\quad g_5 = 112K_2^4 - 144K_2^2 K_3 + 32K_2 K_4 + 18K_3^2 - 4K_5$
$\quad$ etc.

It should be noticed that Eqns. 3:28 and 3:30 are infinite series. However, since Eqns. 3:27 and 3:29 are finite, we must set all the coefficients in the inverted series greater than K_m, i.e., K_{m+1} etc. equal to zero. This is not a point of overwhelming importance, as the coefficients of the inverted series have only been explicitly worked out to 13 terms, i.e., b_{13} (see van Orstrand). Their complexity increases tremendously but in principle it can be extended indefinitely.

THE h AND m COEFFICIENTS

Using Eqn. 3:30, the equation of state (3:14) becomes

$$\frac{P}{RT} = \frac{1}{1 - b/v} [g_1(\sum x C_x) + g_2(\sum x C_x)^2 + g_3(\sum x C_x)^3 + \ldots] \qquad (3:31)$$

Carrying out the indicated division yields

$$\frac{P}{RT} = [g_1(\sum x C_x) + g_2(\sum x C_x)^2 + \ldots][1 + \frac{b}{v} + (\frac{b}{v})^2 + (\frac{b}{v})^3 + \ldots] \qquad (3:32)$$

We must now evaluate b in terms of known quantities. Since, from Eqn. 3:15,*

$$b = \sum b_x N_x \qquad (3:15)$$

then
$$\frac{b}{v} = \sum b_x C_x \qquad (3:33)$$

which, on expansion, becomes

$$\frac{b}{v} = \sum b_x K_x C_1^x = b_1 K_1 C_1 + b_2 K_2 C_1^2 + b_3 K_3 C_1^3 + \ldots \qquad (3:34)$$

---

* We must differentiate carefully between b (without subscript) and b_x (with subscript). b is a variable, while b_x is a molar constant.

Inserting Eqn.3:28 for C_1, expanding and collecting terms, gives

$$\frac{b}{v} = h_1(\sum x C_x) + h_2(\sum x C_x)^2 + h_3(\sum x C_x)^3 + \ldots \qquad (3:35)$$

where the h_x coefficients are

$$h_1 = b_1$$
$$h_2 = K_2(B_2 - 2b_1)$$
$$h_3 = K_3(B_3 - 3b_1) + K_2^2(8b_1 - 4b_2)$$
$$h_4 = K_4(B_4 - 4B_1) + K_2 K_3(30b_1 - 6b_2 - 2b_3) + K_2^3(20b_2 - 40b_1)$$
etc.

Inserting Eqn.3:35 into Eqn.3:32, expanding and collecting terms, gives

$$\frac{P}{RT} = [g_1(\sum x C_x) + g_2(\sum x C_x)^2 + \ldots][1 + m_1(\sum x C_x) + m_2(\sum x C_x)^2$$
$$+ m_3(\sum x C_x)^3 + \ldots] \qquad (3:36)$$

where the m_x coefficients are

$$m_1 = h_1$$
$$m_2 = h_2 + h_1^2$$
$$m_3 = h_3 + 2h_1 h_2 + h_1^3$$
$$m_4 = h_4 + 2h_1 h_3 + 3h_1^2 h_2 + h_2^2 + h_1^4$$
etc.

THE r COEFFICIENTS

Finally, multiplying the two series and collecting terms, we have

$$\frac{P}{RT} = r_1(\sum x C_x) + r_2(\sum x C_x)^2 + r_3(\sum x C_x)^3 + r_4(\sum x C_x)^4 + \ldots \qquad (3:37)$$

where the r coefficients are

$$r_1 = 1$$
$$r_2 = -K_2 + b_1$$
$$r_3 = -2K_3 + K_2(B_2 - 3b_1) + 4K_2^2 + b_1^2$$
$$r_4 = -3K_4 - 20K_2^3 + 18K_2 K_3 + K_2^2(14B_1 - 5B_2) + K_3(B_3 - 5B_1)$$
$$\qquad + K_2(2B_1 B_2 - 5B_1^2) + B_1^3$$
$$r_5 = -4K_5 + 112K_2^4 - 144K_2^2 K_3 + 3K_2 K_4 + 18K_3^2 + K_2^3(28B - 76B)$$
$$\qquad + K_2 K_3(55B_1 - 8B_2 - 3B_3) + K_4(B_4 - 7B_1)$$
$$\qquad + K_3(2B_1 B_3 - 8B_1^2) + K_2(3B_1^2 - 7B_1^3) + B_1^4$$
etc.

Since now
$$\sum x C_x = \frac{w}{M^0 v} \qquad (2:22)$$

Substituting this equation into Eqn.3:37 gives

$$\frac{P}{RT} = r_1\left(\frac{w}{M^0 v}\right) + r_2\left(\frac{w}{M^0 v}\right)^2 + r_3\left(\frac{w}{M^0 v}\right)^3 + r_4\left(\frac{w}{M^0 v}\right)^4 + \ldots \qquad (3:38)$$

multiplying through by v, gives

$$\frac{Pv}{RT} = \left(\frac{w}{M^0}\right) + \frac{r_2}{v}\left(\frac{w}{M^0}\right)^2 + \frac{r_3}{v^2}\left(\frac{w}{M^0}\right)^3 + \frac{r_4}{v^3}\left(\frac{w}{M^0}\right)^4 + \ldots \qquad (3:39)$$

COMPARISON WITH KAMERLINGH ONNES

As can be seen by comparing this with Eqn.1:28, this equation is the virial equation as given in terms of a variable amount of gas (use of w/M^0). The virial equation of Kamerlingh Onnes can be written as

$$\frac{Pv}{RT} = 1 + \frac{B}{v} + \frac{C}{v^2} + \ldots)$$

if we substitute

$$\mathbf{p} = \frac{p}{p_c}$$

$$\mathbf{v}_k = \frac{\mathbf{v}}{K}$$

$$K = \frac{RT_c}{p_c v_c}$$

$$\mathbf{t} = \frac{T}{T_c}$$

then
$$B = R v_c \mathbf{B}$$
$$C = R^2 v_c^2 \mathbf{C}$$
etc.

Kamerlingh Onnes' equation only includes even powers of the volume; if we include the odd powers, we have

$$\frac{Pv}{RT} = 1 + \frac{B}{v} + \frac{C}{v^2} + \frac{D}{v^3} + \frac{E}{v^4} + \ldots \qquad (3:40)$$

which is identical to Eqn.3:39, if the weight of gas chosen is equal to M^0 and

$$B = r_2$$
$$C = r_3$$
$$D = r_4$$
etc.

THE SIGNIFICANCE

The meaning of this is clear. First, the association equation of state, Eqn.3:14, is the closed form of the virial equation of state. Second, since the virial equation of state has been found empirically to be able to describe the properties of gases, the association equation of state will do just as well if the new variables B and C are determined.

In general, only a few virial coefficients (B, C, D, etc.) are required to describe a gas over most of its range*. Therefore it is reasonable to try to determine the various K's and B's, since a knowledge of them will give us the value of C and b/v.

THE FORM OF THE VIRIAL COEFFICIENTS

The determination of the individual K_x's and B_x's must proceed through a knowledge of the virial coefficients. Let us assume that the virial coefficients are known and, hence, that the r's are known. Let us start with the first equation

$$r_2 = B_1 - K_2 \qquad (3:41)$$

We will show that all the other r coefficients have the same form

$$r_3 = -2K_3 + 4K_2^2 + K_2(B_2 - 3B_1) + B_1^2 \qquad (3:42)$$

rewriting

$$r_3 - 4K_2^2 + 3B_1 K_2 - B_1^2 = B_2 K_2 - 2K_3$$

dividing by K_2 gives

$$r_3^* = \frac{r_3 - 4K_2^2 + 3B_1 K_2 - B_1^2}{K_2} = B_2 - \frac{2K_3}{K_2} \qquad (3:43)$$

---

* The reason for this appears in Chapter 5.

Now from Chapter 2 we know that

$$\frac{K_3}{K_2} = K_{1,2}$$

hence $\quad r_3{}^* = B_2 - 2K_{1,2}$ \hfill (3:44)

Similarly

$$r_4{}^* = \frac{r_4 + 20K_2{}^3 - 18K_2K_3 - K_2{}^2(14B_1 - 5B_2) + 5B_1K_3 - K_2(2B_1B_2 - 5B_1{}^2) - B_1{}^3}{K_3}$$

$$r_4{}^* = B_3 - 3\frac{K_4}{K_3} = B - 3K_{1,3} \hfill (3:45)$$

and $\quad r_5{}^* = B_4 - 4K_{1,4}$ \hfill (3:46)

and $\quad r_j{}^* = B_{j-1} - (j-1)K_{1,j-1}$ \hfill (3:47)

It can be seen that these equations can be solved serially, i.e., Eqn.3:41 is solved for B_1 and K_2 (r_2 is known). Using these values and the value for r_3, Eqn.3:44 can be solved for B_2 and $K_{1,2}$, from which we then know K_3. Using then the value for r_4 and the previous values, Eqn.3:45 can be solved, etc. The methods of these solutions are not difficult.

REFERENCES

Boks, J.D.A. and Kamerlingh Onnes, Onnes Communications, No.170a, Leiden, 1924.

Bromwich, T.J.I'A., *Introduction to the Theory of Infinite Series*, Macmillen, London, 1931.

Ginell, R., *Association and the equation of state*, J. Chem. Phys., 23 (1955) 2395-99.

Knopp, K., *Theory and Application of Infinite Series*, Blackie and Son, London, 1928.

Van Orstrand, C.E., *Reversion of power series*, Phil. Mag. [6] 19 (1910) 366-76.

CHAPTER 4

THE NATURE OF THE ASSOCIATED SPECIES

We have spoken of j-mers of varying sizes of j in a general way but have said little about their structure. While the structure of the smaller j-mers is quite simple, the structure of the larger ones is rather complex, so that an examination of their geometry is quite useful and rewarding.

THE FORCES BETWEEN ATOMS

Before we consider the structure we should examine briefly the nature of the forces that hold these structures together. This topic has long been discussed, but often the discussions have been clouded by historical bias. We must therefore make clear in our structure discussions what the implicit assumptions are and how they relate to the historically biased common beliefs.

We believe that there are only two forces existing that bind atoms into aggregates; these are first, an electrostatic force arising from the fact that atoms are composed of positively charged nuclei and negatively charged electrons that are separated in space; and second, an electromagnetic force arising from the fact that these charged subunits (electrons) are in circular motion. The forces existing between particles vary, depending on the magnitude of these two forces and the geometry of the particles. These two forces differ in one fundamental way. The electromagnetic force is directional in character, while the electrostatic force is in general non-directional in character. It is the pattern of combination of these two forces that gives rise to the traditional bond described in chemistry. The simplest way to relate these two concepts is to draw a graph, one axis being strength of electromagnetic force and the other

strength of electrostatic force. On the graph we can plot roughly
the area where covalent forces exist (like C-C bonds in alkanes).
This area would be in the region of high electromagnetic forces.
On the other hand, near the region of high electrostatic forces
would be the realm of electrovalent bonds such as exist in
cesium fluoride. On such a graph one would find that the boundaries of the areas are not distinct and, in fact, in terms of
historical description, certain areas would overlap.

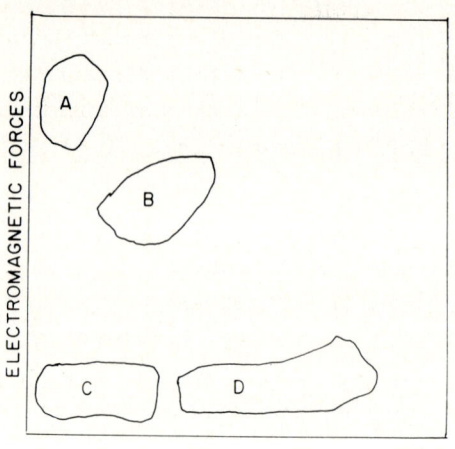

Fig.4.1. Schematic sketch of the classification of bonds according to their
electrostatic and electromagnetic character.
Region of A. covalent bonds; B. ion-dipole bonds; C. rigid dipole-
induced dipole; D. ionic bonds.

In our description of the structure of j-mers we have chosen
the simplest possible case, namely, we assume that the particles
are composed of identical unimers (a 1-mer contains one unimer;
a 2-mer, two unimers; a j-mer, j unimers) and further we assume
that the forces are electrostatic in nature and are spherically
symmetric about the nucleus. This may appear to be a very limiting restriction but apparently it is not. If the system has
electromagnetic bonds, then we consider the unimer to include
the atoms bonded electromagnetically, i.e., we increase the size
of the unimer till it is only bonded to other unimers by

electrostatic forces. Pragmatically, we go about the matter in
another manner. We implicitly assume that we do not know the
particulars of the bonding and calculate the various quantities
for a conventional formula weight of the substance. From the
results we infer what the "true" bonding situation is. Surprisingly, this method gives good results with the substances for
which we have adequate data. Very strong non-symmetrical forces
are generally the result of either dissimilar atoms or strong
electromagnetic forces. Such forces give rise to a rather
strongly bonded particle, which we consider as the unimer of the
system.

SYSTEMATIC DESCRIPTION OF THE j-MER

THE REPRESENTATION OF A BOND AND THE 2-MER

There is only one form of the 2-mer. Historically, this
kind of structure is called the 'dumbell' molecule. However,
this description has implications that are misleading. A dumbell
is two spheres connected by a bar. The bond between two particles
is not a bar in the sense that a bar has rigidity. Sometimes
a representation such as this is valuable for two-dimensional
drawings, but otherwise it should be avoided.

The picture we show in Fig.4.2 is also incomplete and misleading in some of its implications. Unimers (real atoms) are
not little billiard balls waiting to be glued together.

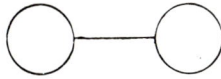

Fig.4.2. Models of the 2-mer.
        Top: stick-and-ball model (dumbell)
        Bottom: tangent model (billiard ball)

What this representation means is this: we consider a unimer to be composed of positive and negative charges, which, when in close proximity to another unimer, give rise to both attractive and repulsive forces. However, these forces do not both follow the same law with respect to their value as a function of distance. That is to say, at any given separation of the unimers, the ratio of the attractive to the repulsive forces is not the same as that at any other separation. Add to this phenomenon the fact that unimers are in motion with respect to one another and hence their separation is continually varying. The result is the dynamic phenomenon, which we sum up by the word "vibration". The particles vibrate around an equilibrium position or distance of separation. In Fig.4.2 we divide the equilibrium distance between the centers of the two unimers by two and draw a circle with this radius around each centerpoint. We now have two tangent circles. This represents two unimers with a bond between them (tangency point). The circle is the region of symmetrical force. This is only an imperfect representation of the abstraction, but using it and thinking of unimers in this sense we can arrive at some useful generalizations.

LINEAR FORMS AND THE 3-MER

There are two forms of the 3-mer (Fig.4.3). The form labeled A is a linear form, while the form B is a close-packed form. Linear forms are unstable forms which have a transitory existence. The orientation or equilibrium point of the component unimers is not specific, geometrically. On the other hand the close-packed forms have a stable geometric orientation or equilibrium point.

THE 4-MER

The variety of forms increases when we consider the 4-mer. In Fig.4.4 are shown all the varieties of the 4-mer structure. There are five linear forms and one close-packed form. The looser linear forms have three bonds. The other linear forms are progressively more tightly bonded. The close-packed form, which is tetrahedral, has six equivalent bonds.

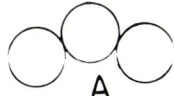

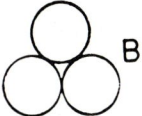

Fig.4.3. Forms of the 3-mer: A - linear form. B - close-packed form.

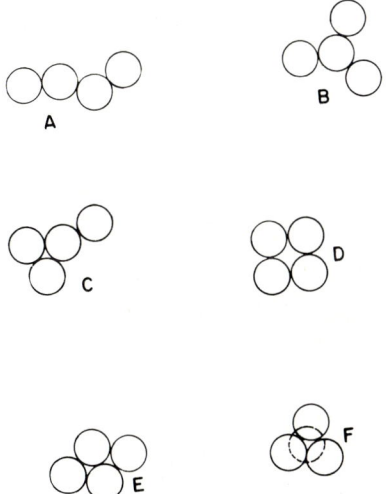

Fig.4.4. Forms of the 4-mer.
    A and B: Strictly linear forms (3 bonds).
    C and D: 4-bond forms. C is formed by the collision of a 1-mer and a close-packed 3-mer. D is an unstable transition form.
    E: 5 bonds, from C or D by making one additional bond.
    F: 6 bonds - tetrahedral form. All four unimers are in 3-holes.

THE 5-MER AND MODEL BUILDING

In Fig.4.5 are shown all the twenty varieties of the 5-mer. The simplest linear forms have four bonds, while the close-packed form has nine bonds. An interesting exercise is the systematic method of finding the orientation and number of possible forms. Let us use the delineation of the 5-mer as an example of the method. In the simple forms this method can be carried out on paper (the 5-mer is about the limit); in the more complex forms the only feasible method is to actually construct three-dimensional models of equally-sized balls.* To form the four-bonded 5-mer we must start with the three-bonded 4-mer, forms 4-A and 4-B. There are two positions where a 1-mer can be added to the form 4-A to give forms 5-I and 5-II. Adding a 1-mer to form 4-B yields forms 5-II and 5-III. This exhausts the possibilities. In Fig.4.6 the relationships of the forms can be seen. We have to exhaust systematically the possibilities of each form but at the same time we have to guard against drawing different representations of the same form and not realizing it. At this point models are useful and, in fact, essential, if forms more complex than five are considered.

THE 6-MER AND TRANSITION BETWEEN FORMS

In the 6-mer there are many more linear forms than in the 5-mer. However, more important than these linear forms are the close-packed ones on which we shall concentrate our attention. From the close-packed 5-mer one can form, by the addition of a 1-mer, the 6a-mer, as shown in Fig.4.7. Since all the 3-hole

---

* It must be emphasized that to be successful in obtaining the geometric relationships, equally-sized spherical balls must be used. I have found table-tennis balls suitable, since they are quite spherical and uniform in size; however, in quantities their cost mounts up and the models become quite large and unwieldy. A more suitable ball I found in 'pearl' beads. These are the base plastic beads that are used to make artificial pearls. They are cast and then ground spherical to close tolerances and are very suitable for models. While they come in many sizes, I have found those from 12 to 15 mm diameter very convenient. They may be obtained in quantity at nominal cost from synthetic-pearl manufacturers.

positions* on the close-packed 5-mer are equivalent, only one form results from this addition. However, six unimers can be arranged in another type of array, which is shown in Fig.4.7 as the 6t-mer. This array, while it cannot be formed directly from the 5cp-mer, can arise from the 5-XIX form, which is a labile form. It can also be formed from the 6a-mer by a transition and this is more likely, as shown in Fig.4.7. While these forms

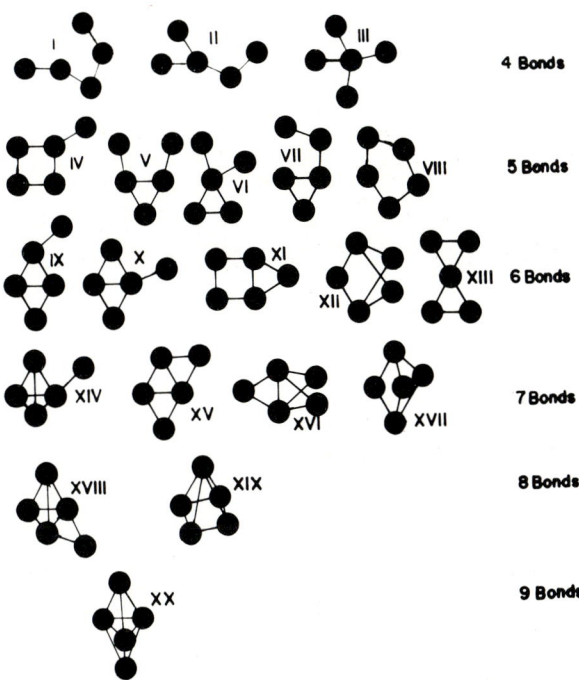

Fig.4.5. Forms of the 5-mer. Some of these forms are drawn in three-dimensional projection. Although all the connected unimers really touch, they are shown with bars (bonds) joining them, in order to make the spatial relationship clearer. (A bar stands for a tangency point.)

---

* A stable position, in which a 1-mer can rest and where it contacts three other unimers, is called a 3-hole; if such a unimer contacts four other unimers it is in a 4-hole position.

look totally different on these bar and sphere diagrams, models show that the amount of movement necessary to accomplish this transition, is slight; single-bond breaks and reformations are undoubtedly simple and easy.

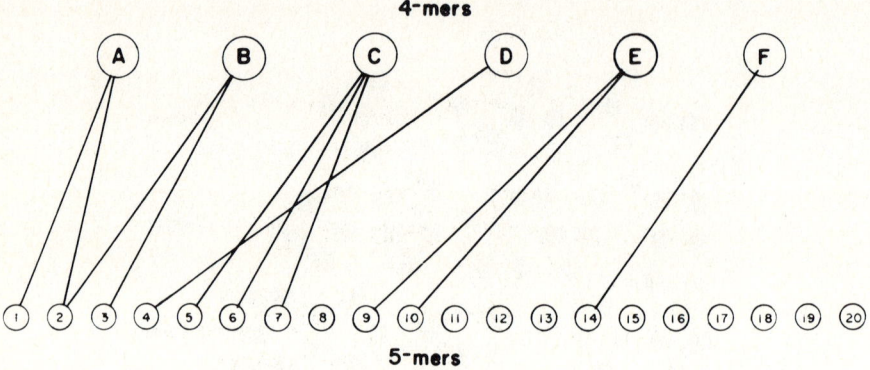

Fig.4.6. Going from the 4-mers to the 5-mers by the addition of a 1-mer and the formation of one bond. The other 5-mer forms are made by interconversions, i.e., formation of more bonds.

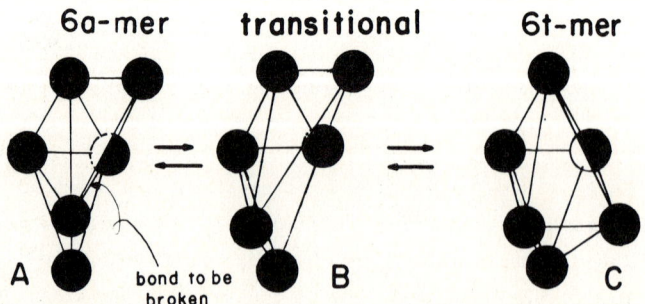

Fig.4.7. The 6-mer transition.
       A - The 6a-mer formed from the 9-bonded (cp) 5-mer and a 1-mer (12 bonds).
       B - The transition form, one bond broken.
       C - The 6t-mer (12 bonds); a bond remade.

THE 7-MER: a *AND* † *FORMS*

The 7-mer is of course more complex than the 6-mer and some interesting features appear with this complexity. Figure 8 shows the positions where additions are possible in the 6†- and 6a-mer. There are 6 stable positions (3-hole) where a 1-mer can add to a 6a-mer and one additional special position. In the 6†-mer there are 8 stable positions where a unimer can add. The resulting forms are shown in Fig.9. While the 6†-mer gives rise to one form (7†-mer), the 6a-mer gives rise to several forms (7a, 7b [*d* and *l*], 7c), one of which appears in right- and left-handed varieties. While the 7b-, 7c- and 7†-mer forms all have 15 bonds,

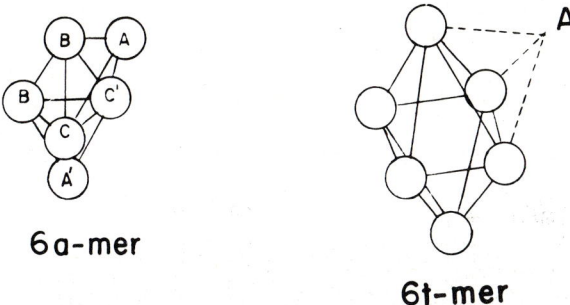

Fig.4.8. Bonding positions in the 6a-mer and the 6†-mer.

    6a-mer - Unimers A and A' are in equivalent 3-hole positions. There are 6 other positions, which are true 3-hole positions (ABC is one of them). ACC'A' is the special position, which makes the 7a-mer. It is not a true 3-hole position, but is closer to a 4-hole position.

    6†-mer - A equals one of the 3-hole positions; there are 8 of these, all equivalent.

the 7a-mer differs, having 16 bonds. In Fig.10 is shown a photograph of a model of the 7a-mer. We can see that the addition of the 7th unimer in the special position (3-hole) results in it coming very close to a 4th unimer. Now, atoms (unimers) are not table-tennis balls and the unimers rearrange slightly, so that 16 bonds are formed, each ever so slightly longer than the usual length. The special bond or "extra bond" in the 7-mer is 10.9% longer (measured on beads) than the other bonds.

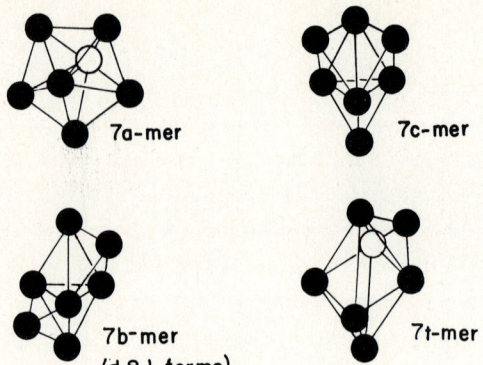

Fig.4.9. Close-packed forms of the 7-mer.
The weak bond in the 7a-mer appears in any one of the 16 bonds or is distributed so that each bond is changed slightly, making all bonds equivalent.
The 7a-, 7b-, and 7c-mers are formed from the 6a-mer, while the 7t-mer comes from the 6t-mer.

Fig.4.10. Photograph of a bead model of a 7a-mer. The weak bond is shown. All the beads touch the other beads. This would be an unadjusted model.

If this distance is distributed over 16 bonds, it means only that the unimers are a little distorted. The consequences of the 7a-mer having 16 bonds is, however, tremendous. The 7t-mer has one unimer in a 3-hole position. This unimer has available to it other 3-hole positions; three that it can reach by breaking one bond and forming another; three that it can reach by repeating this process twice; and one that it can reach by repeating this process three times. The 7b- and 7c-mers have two unimers each in 3-hole positions. These unimers are also mobile, being able to move by breaking one bond and making another. However, the 7a-mer presents another picture. Here there are no unimers in 3-hole positions and, correspondingly, the mobility previously present is missing. The 7a-mer is hence much more stable and is the predominant form in the 7-mer.

THE 8-MER AND HIGHER FORMS

Pursuing this trend of reasoning we can examine the 8-mers and the higher forms. The 8-mer has 15 close-packed forms (Fig.4.11), of which 14 forms have 18 bonds and the 8a-mer has 19 bonds. This kind of discrepancy between the a-mers and the †-mers persists up to much larger sizes. This can be shown by model building.

MODEL BUILDING OF THE HIGHER FORMS

The only way I have found it possible to show some of the features of the higher forms is by model building. Starting with the 7a-mer and 7†-mers I have built up models till the 19a- and the 19†-mers. I have found that the most expeditious way of building models is to make a number of forms at the same time. Since in general it is not possible to add more than one bead at a time (the cement must be given time to dry), it is better to start with a set of forms. For instance, one has a whole series of models, 2, 3, 4, 5, 6a, 7a, 8a-mers and proceeds to add one bead to each model to form the 3, 4, 5, 6a, 7a, 8a, 9a-mers respectively. Each model has "grown" by one bead. The bead that is to be added to each form is placed in the position that yields the most close-packed form. There are often equivalent positions, so that one has a choice. In the case of the †-forms, the choice is of no importance, since whichever way one proceeds, one ends up with the 19†-mer. In the case of the 19a-mer, however, there is not one 19a-mer but a number of equivalent forms, each slightly different in arrangement. Proceeding in

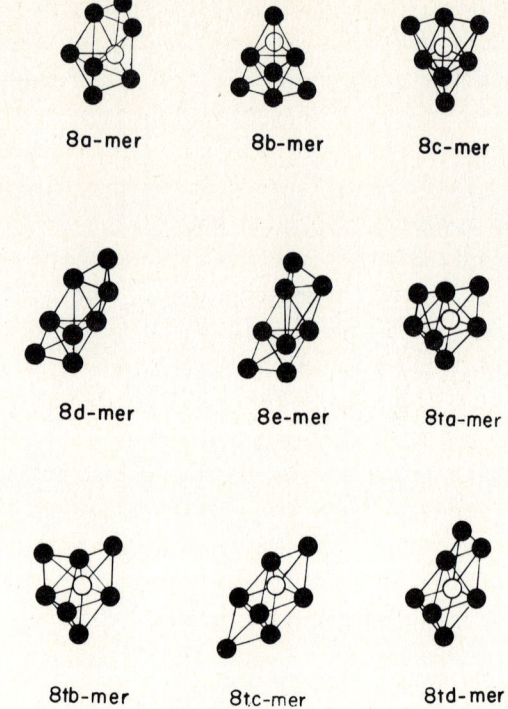

Fig.4.11. Diagram of all the possible close-packed 8-mers. (The white particles are in the back).

this way, one makes only one of the many possible 19a-mers. Thus we obtained the 19a- and 19†-mers as shown in the photograph (Fig.4.12). The a-mer forms persist in having more bonds than the †-mer forms up till 19 and higher* (see Fig.4.13).

THE 19a-MER AND 19†-MER

These two forms are extremely interesting, since the 19†-mer has the outlines of an octahedron, while the 19a-mer looks like a formless amorphous mass with no distinct shape. The 19†-mer intuitively looks like our conceptual picture of what a

---

* Of course, these models are inadequate, for they postulate that unimers have fixed diameters. This is not too serious an assumption. By the very nature of things, models are inadequate and are only symbols; however, they have their value and show certain relationships which are difficult to see otherwise.

Fig.4.12. The 19-mer. A 19a-mer on the left and a 19†-mer on the right. Notice the compact regular array of the 19†-mer. This is typical of crystalline solids. On the other hand, the 19a-mer looks random and has a number of small gaps in its structure. Some are shown. This structure is typically liquid.

micro-micro-grain of a solid should look like, while the 19a-mer appears to be right for a micro-micro-grain of a liquid. The 19†-mer has short- and long-range order, while the 19a-mer has only very short-range order. The 19†-mer is constructed around 6-symmetry. It has a 12 coordination number (CN), i.e., internal unimers have 12 nearest neighbors. The symmetry in the 19†-mer is exact. In contrast, the 19a-mer is arranged in inexact 5-symmetry. Internal unimers in a 19a-mer have various coordination numbers depending on the exact configuration. Most have CN greater than 12, but the distance to these nearest neighbors is not identical (in the 19†-mer the distances are exact).

SYMMETRY: EXACT AND INEXACT, 5- AND 6-

A word needs to be added to define the concept of symmetry as used here. A unimer, in an arrangement usually found in solids, e.g., in simple cubic (sc), body-centered cubic (bcc), face-centered cubic (fcc), or hexagonal close-packed (hcp), is in exact symmetry. That is, each atom has an assigned position in the lattice, from which any movement is impossible, if one thinks of the atom as a rigid billiard ball. In the real

particle the unimers are not rigid billiard balls and, due to
the fact that all the unimers are vibrating, motion is possible.
Any motion away from its location gives rise to large restoring
forces; one might say it is in a deep potential energy valley.

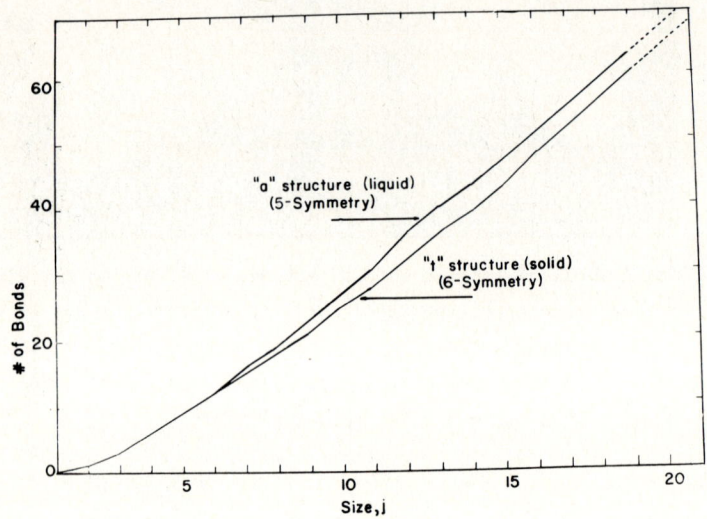

Fig.4.13. J vs. number of bonds. Contact points are considered bonds.
Note the 5-symmetry a-structure has more bonds than the 6-symmetry
†-structure. The curves must cross at some much larger size.
The number of bonds in the †-structure then becomes greater than
that in the a-structure.

On the other hand, a unimer in a liquid is in inexact symmetry. It does not necessarily have an assigned position in a lattice from which any motion is opposed by large forces. On the contrary, it can move (within limits) between several locations with low or no activation energy. It is in a very flat or shallow potential well (Fig.4.14). However, the space it occupies is not enough for two unimers.

Exact symmetry is usually characterized as 4- or 6-symmetry, while inexact symmetry is the result of 5-symmetry. This concept of 6- and 5-symmetry comes in part from the old geometric concepts of what shape of figures can efficiently cover a plane. According to these concepts, a plane can be covered by triangles, squares, or hexagons, but not solely by pentagons. In hcp or fcc the symmetry is 6-symmetry, because the basic arrangement in a plane is six unimers surrounding one. This occurs in the (111) plane. Similar arrangements are found in sc and bcc. However, in inexact symmetry, such as is found in random packing, the preferred arrangement is five.

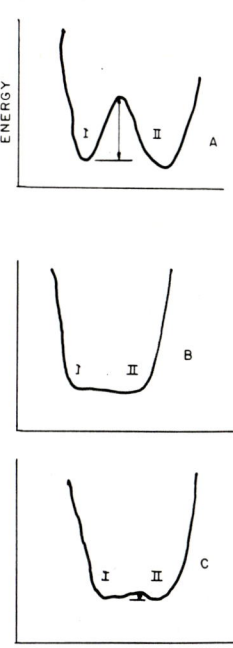

Fig.4.14. Barriers to movement in exact and inexact symmetry positions.
        A - Exact symmetry, potential well. Activation energy barrier between I and II.
        B - Inexact symmetry potential well. No activation energy barrier.
        C - Inexact symmetry potential well. Low activation energy barrier.

BERNAL'S WORK

Bernal (1959) has shown experimentally that when identical spheres are thrown together randomly and then compressed*, the

---

* Bernal used "Plasticene" spheres, which he dusted with talcum and then threw into a rubber balloon. The compression was brought about by evacuating the balloon.

5-sided polygonal face is the most common one to occur on the resulting polyhedra, although polygons with more or less faces also occur. He found the most usual number of faces to be 13, which is the coordination number. All this confirms both C.S. Smith's and L.J. Meijering's work.

The fact that faces with 5-symmetry seem to be a preferred shape in random arrangements, combined with the shape of the 7a-mer, has given rise to the concept of inexact 5-symmetry in the liquid.

Solids with 5-sided faces are, however, known. This is the class of the pentagonal dodecahedron or the pyritohedral class of the isometric system. The distinguishing feature here is exact symmetry. Each unimer is in an exact location, movement from which is opposed by strong restoring forces; hence the necessity of linking the word "inexact" with 5-symmetry.

This work also gives rise to some interesting notes on the concept of nearest neighbors. If a particle is in exact symmetry, i.e., it is a close-packed crystal, then the maximum number of nearest neighbors that it can have is 12 and they are all equidistant from the central unimer when at equilibrium. On the other hand, when a particle is in the liquid and is in inexact 5-symmetry, its packing is not as compact and it can have more than 12 nearest neighbors at distances not quite as precisely fixed. In this way particles in random packing can have more nearest neighbors than particles in exact packing.

REFERENCES

Bernal, J.B., *A geometric approach to the structure of liquids*, Nature, 183 (1959) 141-147.

Bernal, J.B., *The structure of liquids*, Sci. American, 203 (1960) 124-128, 130, 132, 134.

Ginell, R., *Geometric basis of phase change*, J. Chem. Phys., 34 (1961) 992-998.

Meijering, J.L., *Interface area, edge length, and number of vertices in aggregates with random nucleation*, Philips Rev. Rep., 8 (1953) 270-290.

Smith, C.S. and Gutman, L., *Measurement of internal boundaries in three-dimensional structure by random sections*, J. Metals (Jan. 1953) 81-87.

Smith, C.S., *Grain shapes and other metallurgical applications of topology*, paper in *Metal Interfaces*, Am. Soc. Metals (1952) 65-113.

CHAPTER 5

PHASE CHANGE

When beginning a serious theoretical study of the nature of matter, the chief problem has always been how to account rationally for the different phases into which matter separates. Especially exasperating theoretically were the sharp breaks existing at the transition points. Even when proposed theories fitted the properties of the two phases, they nonetheless broke down in the neighborhood of the transition point. The general reason for this breakdown was the fact that the mathematical functions that were used to do the fitting were continuous through the transition point. Further, the models that the mathematics was based on were also continuous and gave no explanation of the nature of the different phases nor a reason or necessity for the transition or break.

With association theory, an explanation evolves and the accompanying mathematics has the requisite break at the transition point. In conventional statistical mechanics the original emphasis is on dynamics, in contrast with association theory where the emphasis is on statics and equilibrium. It might then seem that in considering a dynamic process such as a transition from one phase to another, a theory based on equilibrium, such as the association theory, would be of little help. However, quite the contrary appears to be the case. The static models of association theory are quite helpful in explaining the dynamic process of the transition change. Before we go on to explain these changes, we must consider in more detail the nature of the growth process of the species.

GROWTH AND DEGRADATION

In the previous chapter we spoke of the growth of models and we made the casual statement that the next bead was always placed in the most close-packed position. We now must examine this statement and inquire into what the most close-packed position is and why it should be preferred; at the same time we must look at the transition steps leading to this decision. In particular, we must inquire into the modifications in concepts necessary in going from a model to a real particle. As a first approximation we will consider the atoms to have a fixed radius.

COLLISION OF A 1-MER WITH A 2-MER

In Fig.5.1, view A, for simplicity we consider: (1) the axis of coordinates to be on the 2-mer, i.e., the 2-mer is at rest with respect to these axes, and (2) a 1-mer to be approaching it in a straight line. View B is taken at the moment of impact; ball 3, the impacting unimer, imparts some of its momentum to ball 2 and sets it into motion. However, since ball 3 had components of motion in directions other than the line of motion of ball 2, it moves toward ball 1, which remains attached to ball 2. We must realize at this moment that we are talking about atoms and in addition to the property of impenetrability that atoms have in common with balls (beads), they also have attractive forces that make them stay attached. Now, owing to the impact ball 2 and ball 3 move but only little force has been exerted on ball 1. When ball 3 contacts ball 1, a closely packed 3-mer is formed. The energy of the original impact is now divided between the energy imparted to ball 2 and the energy of impact with ball 1. The exact division is dependent on the angle of impact. Depending on this division of the energy, at least one of three cases occurs. *Case 1:* the energy is more or less equally divided. Here the cp* 3-mer maintains its identity once formed and the energy of impact is converted to rotational energy. The speed of rotation depends on the magnitude of the original impact energy and its relative division. *Case 2:* much more energy has been imparted to ball 2 than is left in ball 3

---

* cp = close-packed

(to impart to 1). The cp 3-mer revolves part of a revolution and the bond 2-3 breaks and ball 2 moves but, being constrained by its attachment to ball 1, revolves around 1 and strikes ball 3 again. Then the cycle repeats itself, i.e., the 2-3 bond breaks and ball 3 revolves around 1 to strike 2, etc. In effect, the cp 3-mer is going to a linear 3-mer, which is going to a cp 3-mer, etc. This also is a kind of cyclic rotation.

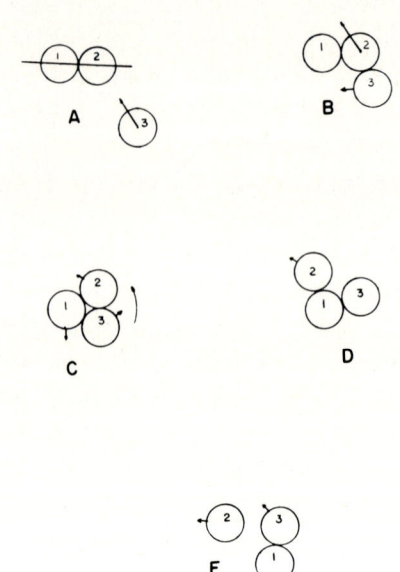

Fig.5.1. The anatomy of a collision.
    A - Before impact: The approach of an impacting particle.
    B - First contact: The transitory linear 3-mer forms.
    C - Second bond forms: The cp 3-mer may be stable and continues rotating.
    D - First step in the degradation: May go from this form to C or E. If to C, it will cycle between C and D.
    E - Final step in degradation: The 2-mer may be left revolving.

Case 3: here the original impact energy is high and/or the division is very uneven. In any case, after bond 2-3 is broken as in case 2, the bond between 2-1 is broken and unimer 2, now a

1-mer, moves away from the remaining 2-mer. This 2-mer again, depending on the division of the energies, may be left rotating.

Another set of cases is also possible, if the energy of collision is several orders of magnitude higher, as in the collisions of a particle emitted from a nucleus, e.g., alpha particle, but this possibility is not of interest here, where we are concerned with thermal collisions.

THE 1-MER AND THE 3-MER OR THE 3-HOLE PROCESS

The collision of the 1-mer and the 3-mer is in a way more complex (see Fig.5.2), and in this section we shall consider these complications. In view A we see a 1-mer approaching a cp 3-mer; in view B one bond has formed and some of the momentum that has been imparted to unimer 2 is transferred to unimer 1. In view C, the second bond has been formed and, since the movement is in 3 dimensions, we see in perspective unimer 4 in contact with unimer 2 and 3 moving toward unimer 1. View D shows the third bond 1-4 formed, yielding a cp 4-mer. This is a stable form, because unimer 4 has fallen into a 3-hole. Since the whole 4-mer is symmetrical, each unimer is in a potential well (3-hole) and to break a bond needs activation energy. At this point the 4-mer is rotating, if it is free.

The degradation steps are shown in view E, F, and G. View E shows the first bond broken. It involves lifting unimer 3 from the 3-hole. Of course, if not too much relative kinetic energy is contained in unimer 3, it will roll over the bond between 2-4 and fall into another 3-hole on the opposite side. (A 3-mer has two sides, each posessing a 3-hole position.) If unimer 3 has enough energy, then another bond is broken, as in view F, and finally a third bond breaks and the unimer is a free 1-mer, as in view G. Here the degradation process is the exact inverse of the growth process; each consists of three steps and each step is alike, i.e., making or breaking one bond. This process is important, because in the larger j-mers the external surface of certain planes is often composed of unimers in 3-hole array. The attachment of a particle to the surface then follows this process.

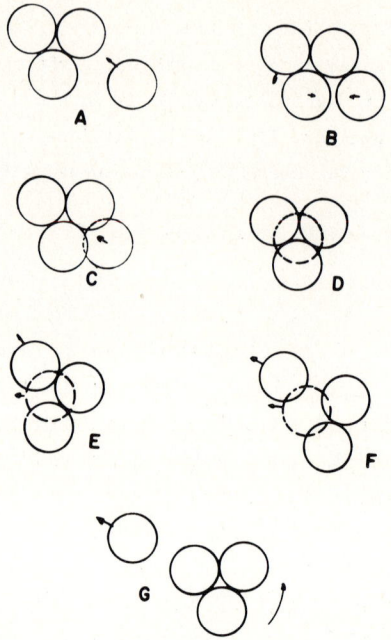

Fig.5.2. Bond formation in a 3-hole.
A - approach, B - first bond forms, C - second bond forms, D - third bond forms, E - first bond breaks, F - second bond breaks, G - third bond breaks (departure).
Process may stop at any of several stages. Transfer of kinetic momentum to rotational momentum most likely. Most stable structure (least energy) is D. The stages traversed and the amount of energy transferred, depends largely on the quantity of energy in the collision and the exact geometry of the collision.

THE 4-HOLE PROCESS

While the 3-hole array occurs very often on the surface of a larger j-mer, it is not the sole arrangement. Often 4-holes are found. Such an arrangement has some interesting properties with profound consequences. In Fig.5.3, view A, is shown the approach of a 1-mer to a 4-hole before impact. View B is the formation of the first bond, while C is the formation of the second bond; view D shows subsequent motion, while in view E the unimer number 5 has fallen into the 4-hole forming two bonds simultaneously. The simultaneity of the formation of these two bonds is because of geometric necessity. It is impossible to form these two bonds sequentially. Now, contrary to the case of the 3-hole process, the removal of a unimer is not the inverse of the accretion. The first step in the removal consists of breaking two bonds simultaneously followed by two single bond breaks. The fact that the removal of a particle from a 4-hole requires as a first step the breaking of two bonds, while the removal of a particle from a 3-hole requires as a first step the breaking of the single bond, makes the unimer in a 4-hole much more stable than the unimer in a 3-hole. This simple fact is of immense importance and will be used to explain many curious phenomena.

THE 5-HOLE AND HIGHER PROCESSES

As particles get more and more complex, 5-hole, 6-hole and higher positions occur. These are of increasing stability. The ultimate position of stability occurs when a unimer is completely surrounded by other unimers. Then it cannot be removed before other processes occur.

THE TRANSITION OF THE GAS TO THE LIQUID

We can now begin to describe and understand the transition of a gas to a liquid. In the process we shall be able to give a clear delineation of the nature of the liquid state.

THE SIZE OF m

According to the derivation of the gas law carried out previously, a gas can be considered to consist of a mixture of species, the 1-mer, 2-mer, 3-mer, etc., all in equilibrium with one another, the limiting size being the m-mer. Although m may be a large number, it is a finite number. This is obvious since the size of the sample is never infinite. Another limitation on the size of m is the temperature or the average kinetic energy of collision. The reason for this is that the higher the temperature, the greater the average velocity of the particle. This implies that at higher temperatures the collisions are more energetic and the lifetime of a specific compound-particle is shorter. The lifetime of 1-mers of a particular species is the result of balance between two factors, the rate of aggregation and the rate of degradation. While the rate of degradation is a function of the temperature, being dependent on the number and violence of the collisions, the rate of aggregation is dependent on the attractive forces between unimers. These forces depend on the internal geometry of the unimer and are probably almost entirely independent of the average velocity of the unimers. If the lifetimes of particles are shorter, then fewer of the larger species form. In other words, the value of *m* decreases with increasing temperature and vice versa.

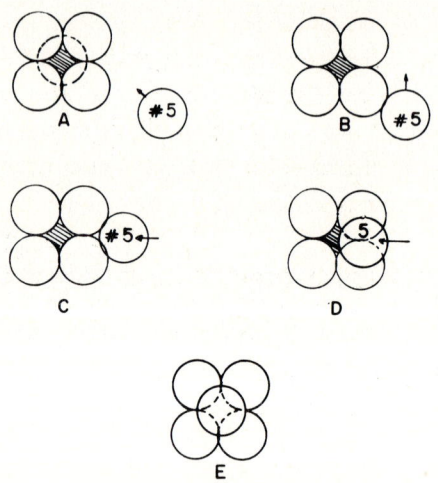

Fig.5.3. Formation of bonds in a 4-hole.
   A - approach (the dotted atom is below the plane of the 4 atoms
    and fixes these 4 in position. It is only partially shown in
    the other views).
   B - first bond forms between atom No.5 and one unimer of the 4-hole.
   C - second bond forms between atom No.5 and the adjoining unimer.
   D - guided movement of unimer No.5 on the unimers of the 4-hole.
   E - simultaneous formation of bonds 3 and 4 between unimer No.5
    and 4-hole. These 2 bonds must be formed simultaneously.
  Degradation of unimer No.5 in a 4-hole.
  The first step is the reverse of step E; the simultaneous breaking of 2 bonds. The next two steps are the reverse of steps C and B.

AT HIGHER TEMPERATURES

Let us now choose a temperature for which m is small, let us say 3. At this temperature we have 1-mers in the system, 2-mers and 3-mers, both linear and cp, i.e., we have binary and ternary collisions occurring and also some internal conversions (cp 3-mers). Most of the 3-mers have the linear form. Since the temperature is so high, most of the bonds that are formed are quickly broken and relatively few cp 3-mers have time to form.

As the temperature is lowered, the violence of the collisions moderates somewhat and the energy content or the amount of transfer of energy that occurs during collisions, decreases. Since the adhesiveness of the bonds is not a function of the average kinetic energy of the particle (the temperature is a function of the average kinetic energy), the lifetime of all the species increases and hence the more complex processes or species have time to occur. Consequently, more complex forms appear or, stated differently, higher order collisions occur. The cp 3-mer occurs more frequently and the various 4-mer species appear.

AT LOWER TEMPERATURES

As the temperature is decreased further, the more complex 4-mers, especially the cp 4-mer, become more common and linear 5-mers appear. This whole process is continuous and the rate of heat loss has had a constant slope. The loss of heat that is necessary if the temperature is to decrease not only comes from a loss in kinetic energy of motion but also from the formation of the additional bonds holding the more complex species together. At this stage, if one stops removing heat from the system, the temperature stops dropping, and, since the system is at equilibrium, the number of bonds present remains constant. If one resumes removing heat, the temperature again is lowered and more complex species are formed, which implies that more bonds are formed. Each bond formed releases energy to the kinetic energy of the system.

> This is really putting the cart before the horse. When the particles collide, if they lose kinetic energy of motion, either to rotation or to another particle, then the bond that is formed is stable. The kinetic energy is lost at the wall of the system to the cooling medium.

Another way of saying this is in terms of the specific heat. Take the converse process; to raise the temperature one degree requires the imput of a specific quantity of heat. For a very simple gas, all the heat goes to kinetic energy of motion; for an associated gas, part of the heat goes to kinetic energy and part to breaking bonds in the system, producing smaller species. Lowering the temperature produces the inverse of this process.

A NEW PHENOMENON

When the temperature is lowered a little more, another phenomenon comes into view. In the process of forming more complex species, the close-packed 5-mer (the nine-bonded form) appears. This form is, however, different from the less complex species in a very important manner. Whereas the less complex species were simply-bonded, meaning, that any unimer held in one of these species could always be removed by a series of single-bond breaks, this is impossible with the cp 5-mer, which is multiply-bonded. The simpler species, the 2-mer, 3-mer, 4-mer, and the linear 5-mer were all simply-bonded, the most complex bond being a 3-hole bond. Multiply-bonded unimers are, however, in such positions that their removal by a series of sequential single bond breaks is impossible; a multi-bond break is necessary. Hence the forms that contain multiply-bonded unimers are immensely more stable. The multiply-bonded unimer in a 4-hole is an example of these extraordinarily stable structures. As a result of this stability the lifetime of the cp 5-mer is longer. This has a number of important consequences. First, the increased lifetime gives time for more complex species to form. The more complex species are in turn more multiply-bonded and have still longer lives, so that we have an autocatalytic process. Second, the increased lifetime implies that the degradation rate falls. This means that the concentration of the simply-bonded species decreases radically, because as they decompose the 1-mers which break off go to form the more complex species. This process is autocatalytic and stops only when the gas is emptied first of 2-mers, then of 3-mers and 4-mers and the higher species progressively. 1-mers do not disappear, however, since whatever decomposition occurs goes first to produce 1-mers. In other words, with the appearance of the cp 5-mer, an autocatalytic process sets in, producing a system of 1-mers and large particles which we call α-mers.

THE HEAT RELATIONSHIPS

The heat relationships during this autocatalytic process are quite interesting. Originally, the rate of heat loss was a result of the balance between formation and breaking of single bonds. As this autocatalytic process takes over, there is only formation of bonds without degradation. Hence the rate of heat loss changes. In fact, the process occurs at constant temperature, because the heat loss is accompanied by the formation of more complex species and none of the heat comes out of kinetic energy. Thus there is a temperature halt; we have reached the condensation point.

MATHEMATICAL CONSEQUENCES OF THE FORMATION OF A GAP

During the temperature halt a gap in the species is formed, i.e., the C_2, C_3 species - up to some size less than C_α - have a concentration of zero. This has disastrous effects on some of our equations. In Chapter 2 we derived an equation which stated that

$$C_j = K_j C_1^j \qquad (2:42)$$

with $\quad K_j = \tfrac{1}{2} K_{1,1} K_{1,2} K_{1,3} \cdots K_{1,j-1}$

Now, if C_2 equals zero, then, since C_1 is not zero, $K_{1,1}$ must equal zero. This implies that not only is $K_2 = 0$ but all $K_j = 0$, no matter what the size of j. How then can C_α exist? (The α-mer is the smallest of the large species.) According to this equation, all sizes must exist up to some limiting size and none beyond it. There can be no gap here.

The resolution of this dilemma is the realization that relationships in the gas are continuous up to the condensation point, where we have a discontinuity, both physically and mathematically. At this point we must change our definitions. Accordingly, we state:

$$
\begin{aligned}
c_1 &= c_1 \neq 0 & &\} \text{ the 1-mers}
\end{aligned}
$$

$$
\begin{aligned}
c_2 &= K_2 c_1^2 = 0 : K_2 = 0 \\
c_3 &= K_3 c_1^3 = 0 : K_3 = 0 \\
c_4 &= K_4 c_1^4 = 0 : K_4 = 0 \quad \text{the gap} \\
&\vdots \\
c_{\alpha-1} &= K_{\alpha-1} c_1^{\alpha-1} = 0 : K_{\alpha-1} = 0
\end{aligned}
$$

$$
\begin{aligned}
c_\alpha &= K_\alpha c_1^\alpha \neq 0 : K_\alpha \neq 0 \\
c_{\alpha+1} &= K_{\alpha+1} c_1^{\alpha+1} \neq 0 \\
&\vdots \quad \text{the large species} \\
c_m &= K_m c_1^m \neq 0
\end{aligned}
$$

$$
c_{m+1} = K_{m+1} c_1^{m+1} = 0 : K_{m+1} = 0 \quad\}
$$

These changes suffice to define our gap and resolve our dilemma. The relationships are summarized in Table 5.1.

THE GAS AND THE LIQUID

After the halt the liquid consists of large α-mers in equilibrium with 1-mers. The concentration of 1-mers is very small and these 1-mers have a very transitory existance. The situation is characterized by the reaction

$$c_{\alpha+1} \rightleftarrows c_\alpha + c_1$$

In a liquid at a given temperature and pressure the range of α is probably very small, although α itself may be very large. We now have a distinct picture of a gas as consisting of a continuous distribution of j-mers from 1-mers to m-mers. In the range, where liquids can form, the particles in the gas have a maximum size of 5, although the cut-off is not sharp; linear forms, a little larger, may occur.

THE LIQUID-GAS PHENOMENON - BOILING

In the range of temperature, where liquids can exist, the boiling phenomenon is not quite the inverse of the condensation. Let us examine this process: The liquid consists of α-mers and internal 1-mers. It has a vapor pressure that is due to the

Table 5.1

Association Equation for Continuous and Discontinuous Cases

Continuous Case	Discontinuous Case
No gap - all species present	*Gap exists - some species absent*
$C_1 = C_1$	$C_1 = C_1$
$C_x = C_x$; $2 \leq x \leq m$	$C_x = 0$; $2 \leq x < \alpha$
$\quad = 0$; $m < x$	$\quad = C_x$; $\alpha \leq x \leq m$
m is the size of the largest species	$\quad = 0$; $m < x$
$C_x = \dfrac{N_x}{v}$	m is the size of the largest species
N_x = No. of moles of particles of size x in mass w. v = volume	*Note:* The gap need not start at 2 but could start at 3 or 4. This entails only slight modification.
Equilibrium constant (double index)	*Equilibrium constant (double index)*
$C_1 + C_x \rightleftarrows C_{x+1}$	$C_1 + C_x \rightleftarrows C_{x+1}$
$K_{1,x} = \dfrac{C_{x+1}}{C_1 C_x}$; $1 \leq x$	$K_{1,x} = \dfrac{C_{x+1}}{C_1 C_x}$; $\alpha \leq x$
Equilibrium constant (single index)	*Equilibrium constant (single index)*
$K_1 = 1$	$K_1 = 1$
$K_x = \tfrac{1}{2} \prod\limits_{y=1}^{x-1} K_{1,y} = \dfrac{C_x}{C_1^x}$; $2 \leq x \leq m$	$K_x = 0$; $2 \leq x < \alpha$
$K_x = 0$; $m < x$	$K_\alpha = \dfrac{C_\alpha}{C_1^\alpha}$; $\alpha > 2$
	$K_2 = \dfrac{C_2}{2C_1^2} = \tfrac{1}{2} K_{1,1}$; $\alpha = 1$
	$K_x = K_\alpha \prod\limits_{y=\alpha}^{x-1} K_{1,y}$; $\alpha < x \leq m$
	$K_x = 0$; $m < x$
$\sum\limits_{x=1}^{m} C_x = \sum\limits_{1}^{m} K_x C_1^x$	$\sum\limits_{x=1}^{m} C_x = C_1 + \sum\limits_{y=\alpha}^{m} K_y C_1^y$
	Actually this definition is redundant, since considering the conditions on K_x and C_x the continuous definition can be used.

pressure of the 1-mer on the wall. If the total volume of the system exceeds the volume of the liquid, then 1-mers escape from the liquid owing to this pressure and exist above the liquid as a saturated vapor. At low temperatures, where few 1-mers exist and the vapor pressure is low, the vapor is dilute, consisting of 1-mers, which strike the wall and are reflected back into the liquid. At this temperature, if the total volume of the system is increased, the amount of vapor increases at the expense of the liquid. If the volume is increased enough, no liquid α-mers are left. The vapor contains 1-mers, 2-mers, 3-mers, etc., depending on the volume.

If the pressure of the vapor is exactly equal to the pressure of the escaping 1-mers and the volume of the vapor is increased, then we have the phenomenon of boiling and the volume of the vapor increases at the expense of the liquid as long as the temperature remains constant, i.e., we must supply energy for this process.

THE CRITICAL STATE

If, instead of just increasing the volume at constant temperature, we keep the volume fixed and increase the temperature, other events occur. Because we are raising the temperature, α now changes, becoming smaller and smaller. At the same time, the vapor becomes more complex. When α becomes equal approximately to 5, both in the liquid portion and in the vapor portion, the distribution becomes continuous and the gap disappears. This is the critical point.

Strangely enough, just before this point is reached we have a vapor and a liquid. The vapor consists of 1-mers through linear 5-mers (and possibly higher linear forms), while the liquid consists of 1-mers and α-mers, where α is a little greater than 5. Since the 2 phases consist of different species, they have different refractive indices and hence are distinguishable by this method. Further, since the α-mers are denser than the vapor, they are at the bottom of the tube. The refractive index differences disappear at the critical point, i.e., the distribution becomes continuous, but the density difference persists for a while. This is known to occur experimentally. The reason is that, whereas the upper part of the tube contains the former gas, consisting of 1-mers through 5-mers, the lower part contains 1-mers and 5-mers, 6-mers, 7-mers, etc. While the distribution of sizes is continuous through the interface, nonetheless the lower part of the tube is denser. These density differences persist for a while but ultimately disappear owing to vibration and minute temperature differences.

Also, in this critical region strange opalescences occur as the system approaches the critical point. The reason for this phenomenon is that in the gas the particles are much smaller than the wavelength of light; in the liquid α-mers, the particles are larger, while in the critical region the size of the particle is just right to interfere with the light, giving anomalous diffraction. Table 5.2 summarizes some of these relationships.

THE VALUE OF α AT THE CRITICAL POINT

Using the concepts of multiply- and simply-bonded unimers we can give a reasonable description of the critical point. This theory concludes that the critical point occurs when the gap disappears, which is when α is approximately 5. The clinching point for this theory would be to verify experimentally that α is indeed approximately 5. We can do this, proceeding as follows. From the concept of the gap

$$N = N_1 + \sum_{j=\alpha}^{m} N_j \qquad (5:1)$$

where N is the total number of j-mers in the liquid (in units of Avogadro's number). The first question is, how large is the spread of sizes between α and m? It cannot be very large, although its size will vary with the absolute value of α. The α-mers are in equilibrium with each other and hence are competing for the available 1-mers or, to put it another way, the ratio of the rates of aggregation and degradation for the α-mer must be equal; otherwise one of the species would grow at the expense of the others. Or, expressing it differently again, the number of multiply-bonded unimers in each particle must be equal; otherwise the one with the more multiply-bonded unimers would be more stable and grow at the expense of the others. In effect we have

Table 5.2

Summary of changes near Critical Point

Temperature	Liquid Phase		Vapor Phase
$T \ll T_c$	$m \simeq \alpha$		1-mers
	α_1-mers & 1-mers		2-mers
$T < T_c$	α_2-mers & 1-mers		1-mers
	$\alpha_2 < \alpha_1$		2-mers
			3-mers
T just smaller than T_c	α_3-mers & 1-mers		1-mers
	2-mers		2-mers
	(?) 3-mers		3-mers
	$\alpha_3 < \alpha_2$		4-mers
T_c	α-mers		1-mers
Boundary disappears	1-mers		2-mers
Density differences	2-mers		3-mers
exist	3-mers		4-mers
	4-mers		5-mers
	$m > \alpha \simeq 5$		
$T \gg T_c$	One continuous phase		
No density	Fluid state		
differences	1-mers	4-mers	
	2-mers	5-mers	mainly linear forms
	3-mers	6-mers	
		etc.	

a balance, where all the particles are about the same size or very close to the same size. Hence we can simplify Eqn.5:1 by assuming that an average size α-mer exists, and the number of these is N_α. Hence Eqn.5:1 becomes

$$N = N_1 + N_\alpha \qquad (5:2)$$

We also know, that the total number of unimers in the system is

$$N_1 + \alpha N_\alpha = \frac{w}{M^0} \qquad (5:3)$$

where w is the size of the sample and M^0 is the formula weight of the unimers. Taking the size of the sample to be equal to M^0, gives

$$N_1 + \alpha N_\alpha = 1 \tag{5:4}$$

Eliminating N_α between Eqns. 5:2 and 5:4, gives

$$\alpha = \frac{1-N_1}{N-N_1} \tag{5:5}$$

EVALUATION OF N

N can be obtained from the general equation of state $P(v-B) = NRT$ if we can evaluate B. However, this cannot be done unequivocally, but we can set reasonable limits. As a lower limit we take B as equal to v_s, the molar volume of the solid at the triple point. This is a common practice and has been used by many investigators, notably Eyring and his co-workers.

The reasoning behind this assumption is that at this temperature and pressure, where the liquid and solid are in equilibrium, the solid usually has a greater density than the liquid. It is generally accepted that liquids contain holes, which decrease its density, while solids have more order and fewer holes. Qualitatively, these ideas are consistent with association theory. According to this theory, the unimers in the α-mers in liquids are arranged in inexact 5-symmetry, while solids are composed of particles in exact 4- or 6-symmetry. This symmetry difference gives rise to more voids in the liquid particles and, of course, tends to increase their covolume; hence we assume this value as a lower limit. Certainly, the covolume of a liquid is not less than this value. There are some notable exceptions to these ideas, e.g., water. Using this approximation for a lower limit in water, gives some peculiar results. However, we know, that B is not a constant and increases with temperature. How much then does it increase as we go from the triple point to the critical point? Here again, the best that we can do is to set an upper limit. We shall choose to increase v_s proportionately to the specific volume-increase of the liquid.

Then our limits are

$$b = v_s \quad ; \quad \text{lower limit} \tag{5:6}$$

$$b = v_s \left(\frac{v_T}{v_{tp}}\right) \quad ; \quad \text{upper limit}$$

The values of the covolume certainly lie between these two values, probably nearer the lower limit, since it is unlikely that the covolume b, will expand as rapidly as the defect-volume $(v-b)$, which the upper limit implies.

VALUE OF N_1

While we would like to have the exact value, it is at this point unavailable. We know that the 1-mers in the liquid phase are in equilibrium with the vapor phase; hence we shall make a very simple assumption that the 1-mers in the liquid are directly proportional to the vapor pressure of the saturated liquid; i.e.,

$$N_1 = k P_v \tag{5:7}$$

The pressure of a saturated vapor has been represented as

$$\log P = A + B \{f(T)\} \tag{5:8}$$

where $f(T)$ has assumed various forms derived empirically. Combining this equation with Eqn. 5:7, we have

$$\log N_1 = (A + \log k) + B \cdot f(T) \tag{5:9}$$

For an appropriate fit, $f(T)$ can be taken as $1/T$. Hence at $T=\infty$

$$\log (N_1)_\infty = (A + \log k) \tag{5:10}$$

At infinite temperature all bonds are broken and $(N_1)_\infty$ is just equal to w/M^0, i.e., $\log(N_1)_\infty = \log w/M^0$. Since we are taking $w = M^0$, then $(A + \log k) = \log 1 = 0$

Therefore $\log N_1 = B/T$ (5:11)

B can be evaluated from the equation

$$\log P = A + B/T \tag{5:12}$$

by linear least squares using experimental values of $\log P$ and T along the saturation curve. If our assumptions are valid, we can now calculate the approximate value of N_1. This simple form of the equation for P fits fairly well, especially around the critical point, where the agreement with experiment is about 1% for a number of substances for which we made calculations.

Calculating some values of N_1 at the critical point shows that $N_1/N \approx 0.02$ at this point (see Table 5.3).

Table 5.3

Value of ratio N_1/N at critical point

N_1 calculated by Eqn.5:12; N by Eqn.3:14; $B = v_s$

Substance	$(N_1/N)_c$
Argon	0.027
Nitrogen	0.020
Oxygen	0.018
Carbon monoxide	0.018
Water	0.0032

Since these calculations are so approximate, we feel justified in saying that Eqn.5:5 reduces to

$$\alpha = 1/N \qquad (5:13)$$

Table 5.4 shows a comparison for a few substances of the values of α at the critical point, both with and without N_1.

Table 5.4

Values of α by Eqn.5:5 (using N_1) and by Eqn.5:13 (omitting N_1) ($B = v_s$)

Substance	With N_1	Omitting N_1
Nitrogen	5.29	5.21
Oxygen	4.83	4.77
Carbon monoxide	5.07	5.00
Argon	5.28	5.17, 5.13

Table 5.5

Values of α calculated by Eqn.5:15 giving both maximum and minimum values; also for comparison values of $(RT/Pv)_c$

$R = 82.0567$

Substance	T_c °K	P_c atm	v_c cc/FW	v_c cc/FW	α min.val.	α max.val.	$\left[\dfrac{RT}{Pv}\right]_c$	Ref.for Data
Ne	44.4	26.19	45.196	15.1177	4.63	22.6	3.08	h
Ar	150.86	48.34	74.558	24.6112	5.13	26.7	3.44	h, n
	150.86	48.34	74.558	24.98	5.17	29.8	3.44	n
Kr	209.4	54.3	92.1839	29.6178	5.06	25.9	3.43	h
Xe	289.74	57.64	119.364	37.0904	5.01	25.3	3.46	h
p-H$_2$	32.976	12.759	64.1437	26.422	5.62	*	3.31	e, n
O$_2$	154.77	50.14	75.00	21.663	4.77	29.4	3.39	n
	154.77	50.54	75.0117	24.6154	4.99	*	3.35	h, j
N$_2$	150.86	48.34	74.558	24.98	5.21	44.6	3.47	n, a
CO	132.91	34.529	93.458	30.318	5.00	40.7	3.38	g, a
CO$_2$	304.2	72.85	94.04	28.2115	5.21	14.9	3.64	k, f, a
HCN	456.66	53.2	138.60	29.222	6.44	24.8	5.08	c, i, m
H$_2$O	647.3	217.978	55.85	19.6896	6.74	**	4.36	d
C$_2$H$_2$	308.33	61.1	112.8	35.72	5.37	23.9	3.67	l
CH$_4$	190.7	45.8	99.0	30.94	5.02	43.1	3.45	o, b
C$_2$H$_4$	282.6	51.2	127.4	39.08	5.13	34.1	3.56	o, b

* These values are negative, apparently partly due to poor data.
** Water is a special case, since the density of the solid is less than that of the liquid at the triple point.

References to Table 5.5

(a) Clusius,K., Diesbergen,U., and Varde,E., Helv. Chem. Acta, 43 (1960) 2059-63.
(b) Clusius,K., and Weigand,K., Z. Phys. Chem., 46B (1940) 1-37.
(c) Coates,J.E., and Davies,R.H., J. Chem. Soc., [London] (1950) 1194-1199.
(d) Dorsey,N.E., *Properties of Ordinary Water Substance*, Reinhold, New York, 1940.
(e) Forsythe,W.E., *Smithsonian Physical Tables*, 9th rev. edn., Washington, 1954.
(f) Hilsenrath,J., Beckett,C.W., Benedict,W.S., Fano,L., Hoge,H.J., Masi,J.F., Nuttall,R.L., Touloukian,Y.S., and Wooley,H.W., *Tables of Thermodynamic and Transport Properties*, Pergamon Press, New York, 1960.
(g) Hust,J.G., and Stewart,R.B., *Thermodynamic Property Values for Gaseous and Liquid Carbon Monoxide*, Tech. Note 202, Natl. Bur. Standards, U.S.A., Washington, 1963.
(h) Kirk,R.E., and Othmer,D., *Encyclopedia of Chemical Technology*, 2nd edn., Interscience, New York, 1966.
(i) Kobe,K.A., and Lynn, Jr.,R.E., Chem Rev., 52 (1953) 117-236.
(j) Kudchadker,A.P., Alani,G.H., and Zwoliński,B.J., Chem. Rev., 68 (1968) 659-735.
(k) Lange,N.A., *Handbook of Chemistry*, 10th edn., McGraw-Hill, New York, 1967.
(l) Miller, S.A., *Acetylene*, Vol. 1, Academic Press, New York, 1965.
(m) Richter,F., *Beilstein's Handbuch der Organischen Chemie*, 4th edn., Springer Verlag, Berlin, 1958.
(n) Roder,H.M., McCarty,R.D., and Johnson,V.J., *Saturated Liquid Densities of Oxygen, Nitrogen, Argon, and Parahydrogen*, Tech. Note 361, Natl. Bur. Standards, U.S.A., Washington, 1968.
(o) Rossini,F.D., *Selected Values of Physical and Thermodynamic Properties of Hydrocarbons and Related Compounds*, Carnegie Press, Pittsburgh, 1953.

Table 5.5 gives the maximum and minimum values of α calculated by Eqn.5:13.

Interestingly, this calculation of α gives us several insights, the first being that although B is a variable, it probably does not have too large a range in values and its value is not far removed from the value of B of the melting solid. Another observation (which is not new) is that water is quite anomalous in its behavior.

THE COMPRESSIBILITY FACTOR

If we now examine the general equation of state we can see by rearranging that

$$N = \frac{Pv}{RT} - \frac{PB}{RT} \qquad (5:14)$$

or at the critical point

$$\frac{1}{N_c} = \frac{RT_c}{(Pv-PB)_c} \qquad (5:15)$$

$(RT/Pv)_c$ has long been defined as the compressibility factor and is approximately constant for a wide variety of substances. The concordance of the experimental value with the values given by the various equations of state has been used as a measure of the quality of the equations of state. Van der Waals' equation gives a value of 2.67, while Dieterici's equation gives values higher than the experimental ones. However, no descriptive reason why this ratio should be approximately constant has been offered. From our work it seems now clear why it should be constant if P_cB is constant, which it apparently is. As a matter of fact, it appears that $1/N$, i.e., α is a better constant than the compressibility factor. If we set

$$\Theta = \sigma/\overline{N_c} \qquad \rho = \text{standard error}$$
$$\Theta' = \sigma/\overline{(RT/Pv)_c} \qquad \theta = \text{fractional deviation}$$

We can compare these quantities for the list of substances in Table 5.5; for these $\Theta = 0.119$, while $\Theta' = 0.153$. Clearly, from this evidence the term that is most constant is $\alpha = 1/N_c$ and not $(RT/Pv)_c$. When better values of B are known, we shall probably find that α is quite constant.

THE LIQUID-TO-SOLID TRANSITION AND NUCLEATION

The nature of the transition of a liquid to a solid has been the subject of much experimentation and speculation. Experimentally, it has long been known that the transition does not always occur spontaneously under specific reproducible conditions and at times it has been found very difficult to induce crystallization of a liquid; this is especially true when the liquid is carefully purified. In such cases it has been shown that the conversion of the liquid to the solid proceeds extremely rapidly once it has been started by introducing a tiny particle or seed of the solid. Often this crystallization can be induced by a foreign particle or impurity. These experiments have given rise to the theory of nucleation, which postulates that for a liquid to crystallize to a solid, nuclei or centers for the crystallization must be present. When small crystals of the solid or of a particular foreign substance are added, they serve as nuclei around which the growth takes place. This process is called "heterogeneous nucleation".

It is, however, also known that under certain conditions, which are poorly defined, a liquid can be induced to crystallize without the apparent addition of seeds of the solid or of foreign impurities. This process is labeled "homogeneous nucleation" and generates much controversy. It is postulated that somehow the liquid generates its own nuclei around which the crystals form. Exactly what these nuclei are and how they are produced is unknown and mysterious and the whole subject is cause for speculation. In fact, there are some investigators who question the existence of homogeneous nucleation entirely and claim that the accidental addition of foreign bodies from the container or otherwise is the cause of the crystallization. Of course, such a theory is difficult to prove one way or the other and gives rise to much polemic.

SYMMETRY

With the advent of association theory, nucleation and the conversion of the liquid to the solid becomes explainable quite directly. The explanation has its genesis in symmetry. In the liquid the unimers in the particle are arranged in inexact 5-symmetry, while in the crystalline solid they are arranged in exact 6-symmetry (or 4-). The problem then, from an association point of view is, how can a system in inexact 5-symmetry be converted to a system in exact 6-symmetry; especially in the light of the fact that the small a-mers in inexact 5-symmetry have more bonds than the t-mers in exact 6-symmetry? Of course, the unimers in the liquid are in constant motion, continually breaking bonds and re-forming others, but the bond breaks are individual events and the unimer does not have a brain and cannot foretell that, if it foregoes forming a bond now, some large number of steps later the whole system of unimers will have more bonds while in 6-symmetry as a solid. Hence the unimers remain, in general, in 5-symmetry. To rearrange a particle in 6-symmetry demands cooperative action of many unimers and this is a very unlikely event.

HETEROGENEOUS NUCLEATION

The solution to this problem is to introduce a particle of the solid. The conditions under which the liquid is existing are such that the liquid is the metastable form and the solid is the stable form. The solid particle is, of course, more multiply-bonded than the liquid, because it is a very large particle (j is large). Hence the 1-mers, which are in equilibrium with the liquid particles attach themselves to the solid seed and it grows at the expense of the liquid particle. Because the liquid phase is being emptied of 1-mers, all the liquid particles grow smaller. A small liquid particle in inexact 5-symmetry is more likely to rearrange itself into 6-symmetry because less cooperation is needed. Hence more nuclei form. This is an autocatalytic process and the liquid crystallizes rapidly. Foreign nuclei, which have a specific structure similar to the solid, may replace the solid seed by tightly or multiply-bonding the 1-mers and thus inducing crystallization.

HOMOGENEOUS NUCLEATION

If the above describes heterogeneous nucleation, how then does a particle whose unimers are in inexact 5-symmetry, convert to exact 6-symmetry without external seeds? Let us examine some stratagems. The classic device of scratching the beaker is obviously a method of introducing particles of glass from the scratch and the scratch surfaces. Sometimes these are of the right orientation and then the crystallization proceeds, but this is really heterogeneous nucleation with foreign seeds.

Another method is to agitate the liquid more or less violently. With very clean water, one can supercool it well below the temperature at which it would normally freeze, if the water is kept still. A supercooled sample of this sort will often crystallize suddenly upon very slight stirring. With other samples under the same conditions, much more violent agitation may be needed. The explanation of these peculiarities lies in the fact that cooperative action of a particular kind must occur. Agitating or stirring mechanically breaks down in part the large inexact 5-symmetry particles; the smaller particles resulting, then can rearrange themselves readily. Since there are many alternative forms of inexact 5-symmetry and only very few of exact 6-symmetry, the probability is that an inexact 5- symmetry form will be the result. But if a few, or even one, exact 6-symmetry particle results, then it acts as a nucleus. Now the process is the same as in heterogeneous nucleation; the only difference is that we cannot be sure that an exact 6-symmetry particle will form. The process is statistical and hence not exactly reproducible.

A similar explanation is useful for the process of storing a liquid sample at dry ice temperatures to induce nuclei formation and then warming it up slowly to bring about crystallization. This method is often accompanied by violent agitation at the low temperature. Here, use is made of the fact that all processes are slowed down at low temperatures, and the hope is that by slowing down reaction rates the rates of formation of nuclei will be enhanced. Pragmatically, we find that such a method is often successful although it is not dependable. The method has a basis in logic, since undoubtedly the temperature coefficients of the multiple-bond process and the simple-bond process differ. A decrease in temperature probably favors the formation of t-forms, because it increases the lifetime of all the forms and hence allows time for more complex j-mers to form. But again, this whole process is problematic and often does not work. However, after the nucleus is formed, warming the liquid allows the crystallization to proceed more rapidly around the nuclei by the normal heterogeneous process. On the whole, homogeneous nucleation is a statistical process, which does occur but not reproducibly. Heterogeneous nucleation is the normal process.

OTHER TRANSITIONS

THE DIRECT GAS-TO-SOLID TRANSITION

This is a very interesting process which usually occurs at much lower temperatures than the gas-liquid transition. Here, the advantages of more bonding enjoyed by a-mers are annulled by the low temperature. Apparently, the low temperature increases the lifetime of all the species. Because the rates of interconversions are slowed down, more of the small †-mers will exist and have a chance to grow and form larger †-mers. Once a large †-form with many multiple bonds is formed, it is stable vis-à-vis the a-mers and thus the gas goes directly to the solid.

SUBLIMATION (solid-gas-solid)

This is quite a simple process. The solid particles, whose unimers are in exact 6-symmetry, are in equilibrium with 1-mers, just as in the liquid phase; hence the vapor pressure of solids. As the total volume of the system is increased, the 1-mers in the solid slowly pass into the vapor phase. A surface at lower temperature, contacting the vapor phase, will cause the gas to revert to the solid. The latter process usually occurs a slight distance removed from the solid sample. The solid, the †-form, forms in preference to the liquid, since the solid is the stable phase at this temperature.

THE MELTING PROCESS

Experimentally, the melting point appears to be rather sharp and it is often used as a calibration point. If one examines the phenomenon carefully, some of this sharpness is blunted. If a crystal is carefully heated, raising the temperature at a slow rate, and the process is observed under the microscope, one sees that the corners melt at a lower temperature than the edges, while the bulk melts still higher. This process is logical from an association point of view.

As one raises the temperature of a solid, its vapor pressure increases. In terms of association theory this means that the α-mers are getting smaller, while the concentration of 1-mers in the solid and hence in the vapor phase is increasing. At the same time, the density is decreasing, i.e., the unimers are oscillating with ever greater amplitude, thus demanding an increased volume in which to vibrate. At the melting point the volume has increased to such an extent that the unimers in a certain region find that they can rearrange themselves into an array with more bonds, i.e., an a-mer arrangement. However, this rearrangement demands more space, so the volume increases. In the process the substance has gone from exact 6-symmetry to inexact 5-symmetry and has melted. The vibrations of the unimers have brought the unimers by chance into an array that has more bonds than the array had previously. Hence the new array is stable compared with the old one, although the new array occupies more space. Normally, in the solid the vibrations of the unimers do not result in such a rearrangement, because the amplitude of vibration is not great enough, and thus the free space for rearrangement is not available.

The fact that the order of melting is corners, edges and bulk is now explicable. The corners have the most free space; hence the melting commences at the corners, progressing to the edges and finally proceeds to the bulk. As the melting proceeds, more and more free space is available because of the increase in volume and hence the process is self-accelerating.

SOLID-STATE TRANSITIONS

Solid state transitions occur for very similar reasons. For instance, a transformation from tetragonal arrangement to an isometric arrangement occurs when the volume of an assymetric molecule or part of a molecule has enough space available for it to rotate in either two or three dimensions. An example of this is ammonium nitrate, which has a number of crystalline modifications in which either the nitrate ion and/or the ammonium ion rotates. The nitrate ion can rotate in two or three dimensions, while the ammonium ion can rotate in 3 dimensions.

SUPERSATURATION IN VAPORS

Another well-known phenomenon that needs explanation is supersaturation in vapors. While superficially the phenomenon resembles the process of the transformation of liquids to solids and the same descriptive terms are used, e.g., nuclei and nucleation, the process is really different. Before we attempt a detailed explanation, let us review some of the experimental facts. Vapor supersaturation is commonly used in the operation of cloud chambers. Here, a volume of liquid and its saturated vapor, cooled to a low temperature, is alternatively expanded and compressed. The first few times it is compressed a fog forms that is allowed to settle into the bulk of the liquid. After the gas in which the vapor is existing has been cleansed, a further compression of the gas or vapor results in the formation of supersaturated vapor, which is very metastable. The passage of charged particles through such a vapor is marked by trails of liquid droplets.

Here the initial compressions and expansions serve to remove the seed nuclei by condensing liquid around them and uniting the droplets thus formed with the bulk of the liquid. The nuclei are usually assumed to be foreign impurities present in the gas. Some of these may indeed be foreign nuclei, but probably the bulk of them are small j-mers in such forms that the addition of further unimers brough about by the compression results in the formation of multiply-bonded α-mers. These pre-liquid particles (nuclei) appear to be quite stable forms and the rapid expansions do not allow time for them to decompose. There is a certain lag in the response of these nuclei to decomposition. Once the gas and vapor have been cleared of these nuclei the rapid expansions and compressions result in a vapor composed mainly of simple linear forms. These may be greater than 5-mers, but because the compression was rapid and the temperature low the rates of transformation are slow. Since there are many more linear forms than multiply-bonded ones, the vapor remains gaseous and is metastable. The introduction of charged particles with an attractive charge results in the acceleration and ionization of the j-mers close to the particle's path and causes collisions. The resulting collisions create multiply-bonded particles, which then act as nuclei for the formation of the droplets.

REFERENCES

Clark, A.L., *Symposium on the critical state: The critical state of pure fluids*, Chem. Review, 23 (1938) 1-15.

See also other papers in this symposium.

Eyring, H., Ree,T. and Hirai,N., *Significant structure in the liquid state I*, Proc. Nat. Acad. Sci. US, 44 (1958) 683.

Ginell, R. and Kirsch, A.S., *Association theory: The discontinuous case and the structure of liquids and solids*, J. Phys. Chem., 74 (1970) 2838-41.

Ginell, R., Kirsch, A.S. and Ginell, A.M., *Association theory and the nature of the critical state*, Ber. Bunsenges. Phys. Chem., 76 (1972) 292-296.

CHAPTER 6

ENTROPY

Classical thermodynamics is the core of modern science. The concepts advanced by this discipline are the abstractions with which we think about the world. Yet this system of thinking has difficulties connected with it; certain of these abstractions are so attenuated that they have acquired a mystique that seems to defy simplification. Further, since thermodynamics is, in essence, a system of differential equations, it defines a set of curves; one must add boundary conditions if one wishes to reach a particular solution. The boundary condition used is in the form of an equation of state. Since the developers of thermodynamics had at their disposal only the ideal equation of state, some unfortunate consequences resulted, which we can mitigate if we use association theory.

THE FIRST LAW

We shall start with some "axiomatic" statements, which are not without their own difficulties. We shall first state that there is a system existing, defining system as a portion of space that possesses boundaries. Ideally, movement of matter and energy across the boundary is controllable by the observer; i.e., a system can be isolated from the world. But in many ways this statement is nonsensical and consists only of words empty of meaning. If we truly isolated a system from the world, then our senses could not perceive it; for example, to see it, light, a form of energy, must enter the system and then be returned to the observer. Hence, if we there were a system, isolated from the world, it would be totally unobservable. We can only postulate such a system, we cannot experience it. Only by being in the system can the observer experience it, but he is in total ignorance of any other system, for nothing can cross the boundaries of an isolated system. The more one thinks of this dilemma, the more one begins to suspect that the very ideas and concepts from which it is constructed are only a result of semantics. The structure of our language forces us to such dilemmas and in reality they do not exist. Let us then put such semantic jargon behind us and be pragmatic about the matter. We shall state that the world is one system, but for experimental purposes we will define a subsystem as a portion of the world that is surrounded by boundaries. Moreover, movement of matter and energy across such boundaries is measurable.

While matter seems a clear concept, energy must be defined. The second axiom is: there is some function of the system, which we shall call energy. At present we use this term without precise definition. We shall also make the stipulation, that the energy in the system is constant. Energy must flow across the boundary for it to increase or decrease. Since these terms are mathematical in nature, we can express them thus: if nothing flows across the boundary, then for the energy, U, it holds that

$$U = \text{constant} \tag{6:1}$$
$$dU = 0 \tag{6:2}$$

These statements are often called the first law of thermodynamics. They are in fact definitions.

> A definition differs from a law thus: a law states that certain relationships exist and may be true or false. If it is true, the relationships really exist, and if it is false, the relationships do not exist, although this point may be difficult to prove. A definition, on the other hand, is a statement of what something is, to put it inelegantly. No question can arise as to whether a definition is true or false; in other words, a definition can be considered to be a label that is attached to certain phenomena or relationships.

These statements, which are called the first law, are in fact statements defining the phenomena; if nothing crosses the boundary, U is constant; anything that crosses the boundary becomes part of U. Use is made of this last statement to give an alternate definition of the first law. We realize that energy exists not only in the system, but also outside of it. Energy of various kinds exist; for instance, work is a type of energy, whether it is mechanical or electrical. However, originally, the kinds of energy that were considered were pressure-volume work and heat. A consideration of the interrelationships of these two types of energy gives rise to the subject of thermodynamics. Further, these two types of energy have some interesting and confusing relationships, so that often the statement is made that the equation

$$\Delta U = q - w \qquad (6:3)$$

represents a statement of the First Law of Thermodynamics. Here w is taken to mean work done by the system. In the differential form, this equation is

$$dU = dq - dw \qquad (6:7)$$

This equation requires that the changes in the energy of the system are due to the heat, q, and the work, w, entering or leaving the system. The equation is merely a reflection of the statement of the law in Eqns. 6:1 and 6:2.

However, as clean and neat as these statements sound, they are really very abstract, and we must put matter into them quite literally; in other words, we must apply them to real substances.

THE SECOND LAW

We will first use a gas, which according to association theory is to be considered to consist of a mixture of species: 1-mers, 2-mers, 3-mers, 4-mers, and 5-mers (linear forms only), and perhaps linear forms of some of the higher species. We also make the assumption, that the amount of matter in the system is limited and constant.

> We know from other considerations, i.e., Einstein's postulation, that matter is really a form of energy. This statement is equivalent to stating that this form of energy does not cross the boundaries of our system. In fact, we shall here ignore the existence of Einstein's relationship entirely, as it is not directly pertinent to our present development.

We now have a very logical question to ask, namely, what is the connection between the function called the energy and the matter in the system? Or to word the question differently, where in the system is this energy found? In general terms, the answer is that the energy of the gas is to be found in the nature of the properties of matter. Every real substance has associated with it a number of properties, which serve for its characterization, i.e., to differentiate it from other substances; such properties are the density, specific volume, pressure, color, etc. An addition of energy results in the alteration of some of these properties. The question of what properties are altered, when the energy is changed, is not new and was asked quite early in the study of thermodynamics but was only answerable in a general way. Only with the advent of the kinetic-molecular hypothesis did more specific answers become available. Empirically, we know that, if we have a gas confined to a cylinder with a piston, we can do mechanical work on the system by pushing down on the piston. The properties that are altered are the volume and the pressure. In earlier days this answer sufficed, but soon the question arose: what is the pressure? From the point of view of empiricists, this is unanswerable. In the 19th century, when the kinetic-molecular hypothesis was promulgated, an explanation for pressure became available. This theory postulated that the gas was composed of particles, which were in violent motion and were striking each other and the walls continuously; hence the pressure of the gas was due to the collision of the particles with the wall of the container. The volume, according to this theory, was the portion of space that the particles in the gas occupied. This derivation yielded the equation

$$Pv = n\overline{mc^2}/3 \qquad (6:5)$$

Boyle's Law came out of this equation quite simply if one assumed that the right-hand side of the equation was constant at constant temperature. This meant, that the right-hand side of the equation was concerned with temperature and the equation could be reconciled with the ideal equation by setting

$$n\overline{mc^2}/3 = RT \qquad (6:6)$$

The interpretation of temperature was now apparent. The temperature was a function of the average squared velocity of the particles. Going back to the piston-and-cylinder example - in doing mechanical work on the system it can be shown that the work done on the system is

$$dw = Pdv \tag{6:7}$$

How does this energy appear in the system, or rather, where is this energy in the system? Eqn.6:5 holds if one considers all the particles identical, of mass, m, and of average velocity, $\overline{c^2}$. From mechanics, the kinetic energy of an average particle is $m\overline{c^2}/2$ and for n identical particles it is

$$U_k = nm\overline{c^2}/2 \tag{6:8}$$

Hence

$$Pv = \tfrac{2}{3}(nm\overline{c^2}/2) = \tfrac{2}{3}U_k = RT \tag{6:9}$$

RELATIONSHIP TO ASSOCIATION THEORY

The difficulties stem from one of the assumptions made in the derivation of Eqn.6:9, namely, the assumption, which is made implicitly, that the collisions in the kinetic-molecular theory are instantaneous in nature. Of course we realize as soon as we examine this statement that instantaneous collision is not a reasonable concept. A more reasonable concept is that collisions occur over a period of time, however short it may be. "Instantaneous" is only a figure of speech, meaning in the common context of the everyday world a period of time so short that it seems not to exist; but we must explicitly recognize it for what it is: a figure of speech. Sometimes this concept of instantaneousness is rationalized by using a statistical argument. The statement is made that there are an enormous number of particles striking the wall and that at any instant the number striking the wall and the number leaving the wall, with identical properties, is equal. This is a perfectly valid concept, but it is not equivalent to instantaneous collision and, further, this argument leads to other conclusions. If the collisions are not really instantaneous, but the concept is only statistical, then this means that the particles, during the interval when they are not in motion, are joined to the wall. But the wall as a smooth featureless surface is also not a tenable concept, especially when we are discussing events at the atomic level; the wall consists of atoms, and a collision with the wall means a collision with the atoms in the wall. Thus it is not unreasonable to say that when a collision occurs away from the wall in the body of the gas it is instantaneous in the same sense as a wall-collision. This leads inescapably to the conclusion that collisions in the gas yield 2-mers, 3-mers, etc. of unknown lifetimes, which is the exact assumption of association theory.

THE TRANSITION STATE

This concept has a considerable impact on the notions we have about the energy relationships in the gas. According to classical ideas, the gas molecule has a momentum of *mu* before collision and a momentum of *-mu* after collision, which is considered to be instantaneous. If the collision is not instantaneous, then a period of time must elapse during which the momentum is zero. This means that, whereas the particle had kinetic energy both before and after the collision, during the collision the kinetic energy must be converted into some sort of potential energy. To understand the process more thoroughly, we must examine it much more carefully.

When the two particles collide, the resulting mass is, of course, the sum of the two individual masses; but, more important, the resulting energy is the sum of the two individual energies. This energy is no longer entirely kinetic energy, but most of it is stored as a stress in the electronic fields in the particles. The combined particle is "hot", i.e., if the stress energy it now has were converted into kinetic energy, its energy content would be greater than that of the average particle. The particle cannot, however, store this stress energy but must reconvert it to kinetic energy. The analogy here is to the behavior of real billiard balls. When two billiard balls collide, their shape is deformed during the collision, building up in the billiard ball a deformation stress. This stress is a potential energy. When all the kinetic energy of motion is converted to a potential energy of stress, then the deformation stress is relieved by the billiard balls reconverting the stress to kinetic energy; in other words, snapping back to a sphere. In the particles probably something very similar happens, the stress being in the shape of the electronic charge field of the particle. If the stress is relieved by being reconverted into kinetic energy of motion, then the collision is an elastic one. If the stress energy is passed on to another particle due to a third-body collision, then there is insufficient stress energy remaining in the original double particle to allow it to split apart against the ever-present forces of attraction; thus a stable 2-mer has formed and the system has now extra kinetic energy. Consequently, since the temperature constancy is dependent on the average kinetic energy remaining constant, the temperature must have risen. If this is an isothermal system, the extra energy is transferred to the external reservoir.

QUANTITATIVE RELATIONSHIPS

We can express these concepts quantitatively by proceeding thus. Let us imagine that we have a gas, which is a system containing a mixture of species in equilibrium from which we are slowly removing energy, so that the system remains in equilibrium. Since we have set up a temperature gradient between the system and its surroundings, and thus are removing heat from the system, the average velocity of the particles will decrease. However, not all the energy transferred from the system will come from the velocity decrease; there are at least two other sources of this energy. The first of these is the containment energy; and the second is the energy change due to the decrease in the number of particles. The containment energy is a concept that we usually consider in a very superficial way. It is, of course, of prime importance in the gas phase, but of lesser importance in the liquid and solid phases. A gas placed in a vacuum will, in the absence of other forces, expand without limit to fill the void.

In gases at constant pressure the decrease in temperature is accompanied by a decrease in the volume owing to the decrease in the number of collisions with the wall. This decrease in volume gives rise to the Pdv term. The P here is P_e, the external pressure. The term is much smaller in liquids and solids, because the decrease in volume with a decrease in temperature is much smaller. The second factor mentioned above is, however, of great importance in all three phases. Since all the species in the system are in equilibrium, if one changes the condition under which they exist (i.e. temperature), the total number of particles changes and also the number of each species present changes.

Nonetheless, the mass of the sample is constant, and hence the total number of unimers remains fixed; i.e., $\sum x N_x$ = constant, where N_x is the number of moles of the species with the degree of association, x, in the sample. It may easily be seen that energy changes ensue when the distribution of species changes. Since we have a system at equilibrium containing a distribution of sizes, there is an average energy; it can be considered as an abstraction that every particle possesses this energy. Under such conditions, let us allow two particles to combine to form a third. The resulting particle has twice the average energy and thus is "hot", since its energy is the sum of the energies of the forming particles; hence the average energy of the system has increased. If the temperature of the system is to remain constant, then either the extra energy must be removed or a compensating change must occur somewhere in the system.

According to the Ideal Gas Law, Eqn.6:9, the kinetic energy of an ideal gas - one composed wholly of 1-mers - is equal to $\tfrac{3}{2}RT$. This is true for an Avogadro number, N_0, of 1-mers. An average 1-mer would have an energy of

$$\tfrac{3}{2}kT = \tfrac{3}{2}(R/N_0)T \qquad (6:10)$$

For an ideal gas this is the whole energy; it is all kinetic energy. From previous discussions we know that a gas composed of 1-mers is not a tenable concept; therefore we must refer to the equation of state of the associated gas, namely

$$P(v - B) = N_w RT \sum N_x \qquad (6:11)$$

where $\sum N_x$ is the number of all species and B, the covolume, a variable, is given by

$$B = B_1 N_1 + B_2 N_2 + B_3 N_3 + \ldots + B_m N_m \qquad (6:12)$$

$N_w = w/M_0$, where w equals the weight of the sample and M_0 equals the formula weight of the unimer. The average kinetic energy per particle is still $\tfrac{3}{2}kT$, and hence the total energy of the substance is

$$\Delta U = \int dU = \int d(\tfrac{3}{2}RTN_w \sum N_x) - \int P d(v - B) \qquad (6:13)$$

Here the first term on the right-hand side of the equation is the kinetic energy of motion and the second term is the work put into the system, $(v-B)$ being the change in the free volume in the gas.

In classical thermodynamics, the equation

$$dU = TdS - Pdv \qquad (6:14)$$

is used. Here v must be interpreted as meaning the free volume, $(v-B)$; hence, if we equate Eqn.6:13 and the modified Eqn.6:14, we have

$$\int TdS = \int d\tfrac{3}{2}RT\sum N_x \qquad (6:15)$$

if we take N_w to be 1 (i.e., one formula weight of unimers is the quantity taken). For a process where the temperature is constant, Eqn.6:15 can be integrated and becomes

$$\Delta S = \tfrac{3}{2}R\Delta\sum N_x \qquad (6:16)$$

A constant temperature process such as this would be a change of state. During such a process the meaning of entropy, ΔS, becomes especially clear, it being equal to a constant times the change in $\sum N_x$. The meaning is further clarified if one considers Eqn. 6:15. This equation implies that the change in quantity of energy called the heat is equal to the change in the temperature (average velocity of the particles) times the number of such particles. This seems to be a reasonable concept.

THE DILEMMA OF $\Delta\sum N_x$

We are now in the position to do some calculations and see how these concepts look mathematically. In Table 6.1 are shown some values for $\Delta\sum N_x$ calculated for the liquid-gas transition. As can be seen by examining the values for $\Delta\sum N_x$, something seems wrong. How can it be, if we start with only one mole of unimers, that the change in the number of particles, $\Delta\sum N_x$, is greater than one? Whatever this quantity is, it is obviously not the change in the number of particles. Tracing back the chain of derivations we see that the use of this term started with the derivation of the equation of state, and hence we must reexamine this derivation.

Table 6.1

Values at liquid-gas transition

Substance	State	Temperature °K	Pressure atm	$\Delta \sum N_x$
N_2	triple pt.	63.148	0.1237	7.658
N_2	boiling pt.	77.347	1.00	5.775
Ar	triple pt.	83.800	0.680	6.290
He	boiling pt.	4.224	1.00	1.531

THE EQUATION OF STATE REVISITED

In the derivation of the equation of state we have the equation

$$P(v - B) = \sum_x \sum_\phi \tfrac{1}{3} n_x m_x c_\phi^2 \qquad (6:17)$$

This is not quite the form in which it is previously given, but it is an equivalent form. The right-hand side of this equation was worked on to derive Eqn.6:11. The essence of the derivation is that we could substitute an average square energy term for the sum over ϕ in this equation and then make the assumption that the average kinetic energy of each species was the same as any other. This gave us the required equation, but there are certain difficulties with this concept. We recognize that the right-hand side of this equation is a kinetic energy term, and from elementary physics we know that it is $\tfrac{2}{3} U_k$ and, further, that to reconcile it with the ideal gas equation it must be equated to ξRT, where ξ is a factor. The question is: just what is the factor? In the original derivation we postulate that the factor was the number of species. This is a valid assumption if we have a series of linear j-mers. Here the addition of a given minimum amount of energy results in the splitting of a bond and the formation of two particles, where one existed previously. Each successive addition of the same amount of energy results in the same increment in kinetic energy. We can sum the energy by summing the number of species and multiplying by the energy of the 1-mer. This meaning is represented by a term such as $RT \sum N_x$. $\sum N_x$ here should be between 0 and 1 in value if we start with an Avogadro number of unimers. While in dilute gases such a situation does exist, in concentrated gases and in liquids and solids it is not the case. In such cases, the unimers are multiply-bonded; kinetic energy may be added to a particle in the amount necessary to break a bond, but there will not be any additional particles formed. In other words, to sum the amount of energy in the case of multiply-bonded particles the number of broken bonds must be counted. We have to use such an inverse statement, because the maximum in kinetic energy occurs when there are no bonds. The more bonds that form, the less the kinetic energy, until at absolute zero, where there is no kinetic energy, perfect order exists and the maximum of bonding is present.

ENTROPY AND THE SECOND LAW

Hence Eqn.6:15 becomes

$$\int T dS = \tfrac{3}{2} R \Delta T B \qquad (6:18)$$

During a phase change, when the temperature is constant,

$$\Delta S = \tfrac{3}{2} R \Delta B \qquad (6:19)$$

if one substitutes B for $\sum N_x$ (or the factor ξ). The values in Table 6.1 in the column labeled $\Delta \sum N_x$ are now comprehensible, since $\Delta \sum N_x$ is really ΔB, the number of bonds broken. It is now perfectly clear why entropy has been such a mystery and why it was connected with so many difficulties. Entropy, defined as the number of broken bonds times a constant ($\tfrac{3}{2}R$), is a fairly simple concept, but if we deny the existence of bonds but admit the existence of entropy, then entropy becomes a mysterious term. The question remains: what further ramifications does this concept have and where does it lead?

BONDS AND STRUCTURE

One can use the definition for B to gain some insight into the liquid and solid states and especially into the processes occuring during the transition from the solid to the liquid and from the liquid to the gas. From Eqn.6:13 we can write

$$dU = d\tfrac{3}{2} N_w RTB - Pd(v - B) \qquad (6:20)$$

if we substitute B for $\sum N_x$. Since by definition,

$$H = U + P(v - B)$$

writing here $(v - B)$, the free volume, for v, we can write

$$dU = dH - dP(v - B) \qquad (6:21)$$

If we substitute in Eqn.6:21 the equation of state

$$d[P(v - B)] = d(N_w RTB) \qquad (6:22)$$

we have

$$dU = dH - d(N_w RTB) \qquad (6:23)$$

Combining Eqn.6:23 and 6:20, we have

$$dH = d^{5}\!/_{2} N_w RTB - Pd(v - B) \qquad (6:24)$$

At P and T constant, Eqn.6:22 becomes

$$Pd(v - B) = N_w RT dB \qquad (6:25)$$

Substituting Eqn.6:25 into Eqn.6:24 and integrating, we have

$$\Delta H = {}^{3}\!/_{2} N_w RT \Delta B \qquad (6:26)$$

This equation is valid for a phase change where P and T are constant. In Table 6.2 we have calculated ΔB for a variety of substances at both the boiling point and the freezing point. This is an easy task, since the values of these transition points, i.e., ΔH and the temperature, are commonly determined quantities and have been compiled. The insights offered by Table 6.2 are most interesting. The first observation is that ΔB_v, the number of bonds broken per mole of unimers vaporized, falls into a narrow range of values for many substances of a very diverse nature. The values of ΔB_v for water = 8.7360, for ethyl alcohol = 8.7949 and for the metal gold = 8.4884. Many other substances have comparable values. While there are some resemblances between water and alcohol, it stretches the point to say that they resemble gold; still, the values are very close to one another. The same conclusion can be drawn for the values of ΔB_f, the corresponding quantity for the fusion process. This general conclusion is not too surprising, since it has long been known that

Table 6.2

ΔB_V, number of bonds broken at vaporization and ΔB_f, number of bonds broken during fusion for a variety of substances.*

	Liquid → Gas				Solid → Liquid			
Substance	$T°K$	P mm	ΔH_V kc/mol	ΔB_V	$T°K$	P mm	ΔH_f kc/mol	ΔB_f
O_2	90.19	760	1.630	6.0632	54.40	1.1	0.106	0.6537
O_3	162.65	760	2.59	5.3422				
H_2	20.39	760	0.216	3.5539	13.96†	54.0	0.028	0.6729
H_2O	373.16	760	9.717	8.7360	273.16	760.0	1.4363	1.7641
He	4.216	760	0.020	1.5915				
Ne	24.57	324	0.431	5.8850	24.57†	324.0	0.080	1.0923
Ar	87.29	760	1.558	5.9879	83.85†	516.5	0.281	1.1243
Kr	119.9	760	2.158	6.0382	115.95†	549.0	0.391	1.1313
Xe	165.1	760	3.021	6.1387	161.3	611.0	0.549	1.1419
Rn	211	760	3.92	6.2327	202		0.693	1.1509
F_2	85.24	760	1.51	5.9430	55.20		0.372	2.2609
HF	293.1	760	1.8	2.0603	190.09		1.094	1.9308
Cl_2	239.10	760	4.878	6.8444	172.16		1.531	2.9834
HCl	188.11	760	3.86	6.8841	158.94		0.476	1.0047
Br_2	298.16	214	7.34	8.2589	265.9		2.52	3.1795
HBr	206.4	760	4.21	6.8420	186.28†		0.575	1.0356
I_2					386.8		3.74	3.2438
HI	237.80	760	4.724	6.6646	222.36		0.686	1.0350
S					392		0.293	0.2508
SO_2	263.14	760	5.955	7.5922	197.68†		1.769	3.0022
SO_3	316.5	760	9.99	10.5893				
H_2S	212.82	760	4.463	7.0354	187.63†	173.9	0.568	1.0174
H_2SO_4					283.51		2.36	2.7927
$H_2SO_4 \cdot H_2O$					281.69		4.63	5.5142
SF_4	233	760	5.20	7.4872				
SF_6	222.5†	1700	4.08	6.1518	222.5†	1700	1.20	1.8094

Table 6.2 continued

Substance	Liquid → Gas				Solid → Liquid			
	$T°K$	P mm	ΔH_v kc/mol	ΔB_v	$T°K$	P mm	ΔH_f kc/mol	ΔB_f
N_2	77.34	760	1.333	5.7823	63.15†	94	0.172	0.9138
NO	121.39	760	3.293	9.1009	109.51†	164.4	0.550	1.0685
N_2O	184.68	760	3.956	7.1864	182.39†	658.9	1.563	2.8764
N_2O_3	275	760	9.4	11.4679				
N_2O_4	294.31	760	9.11	10.3845	261.96†	139.78	3.502	4.4849
NH_3	239.73	760	5.581	7.8102	195.40†	45.57	1.351	2.3196
HNO_3	293	48	9.43	10.7974	231.56		2.503	3.6264
$HNO_3 \cdot H_2O$					235.53		4.184	5.9596
$HNO_3 \cdot 3H_2O$					254.69		6.954	9.1600
CO	81.66	760	1.444	5.9324	68.10†	115.3	0.200	0.9853
CO_2					217.0		1.99	3.0766
CH_4	111.67	760	1.955	5.8733	96.8 †	87.7	0.225	0.7798
HCHO	253.9	760	5.85	7.7298				
HCOOH	373.7	760	5.32	4.7760	281.46†	18	3.03	3.6116
CH_3OH	337.9	760	8.43	8.3698	175.28		0.757	1.4489
CF_4	145.14	760	3.01	6.9575	89.47		0.167	0.6262
CH_3F	195.1	760	4.23	7.2737				
CCl_4	349.9	760	7.17	9.6256	250.3		0.60	0.8042
CH_3Cl	248.94	760	5.15	6.9404	175.44†	65.68	1.537	2.9391
CH_2Cl_2	313	760	6.69	7.1706	176		1.10	2.0968
$CHCl_3$	334.4	760	7.02	7.0428	209.7		2.2	3.5196
CS_2	319.41	760	6.4	6.7221	161.1		1.05	2.1991
HCN	298.86	760	6.027	6.7656	259.92	140.4	2.009	2.5931
C_2H_2	191.7	900	4.2	7.3502	191.7 †	900	0.9	1.5730
C_2H_4	169.45	760	3.237	6.4088	103.97†	0.9	0.8008	2.5840
C_2H_6	184.5	760	3.517	6.3944	89.89†	0.006	0.6834	2.5506
CH_3COOH	391.4	760	5.83	4.9971	289.77	760	2.80	3.2417
CH_3OCH_3	248.34	760	5.141	6.9450	131.66		1.18	3.0068
C_2H_5OH	351.7	760	9.22	8.7949	158.6		1.20	2.5384
$HCOOCH_3$	304.7	760	6.75	7.4320	174		1.6	3.0649

Table 6.2 continued

	Liquid → Gas				Solid → Liquid			
Substance	$T°K$	P mm	ΔH_v kc/mol	ΔB_v	$T°K$	P mm	ΔH_f kc/mol	ΔB_f
CdSb					729		7.66	3.5251
$Cd(NO_3)_2 \cdot 4H_2O$					332.7		7.8	7.8653
$Hg(C_2H_5)_2$	432	760	10.1	7.8435				
Cu	2855	760	72.8	8.5546	1355.2		3.11	0.7693
Cu_2O					1502		13.4	2.9930
CuCl					703		2.4	1.1453
Ag	2466	760	60.72	8.2606	1234.0		2.70	0.7340
Au	2933	760	74.21	8.4884	1336.16		3.03	0.7608
AuSn					691		6.12	2.9713
AuCd					900		4.28	1.5954
Pt					2042.5		5.2	0.8541

* Values of ΔH_v and ΔH_f from Rossini, F.D. et.al.
† Values taken at triple point

substances in corresponding states, e.g., boiling point, etc., have similar properties. These resemblances are generally formulated as the Law of Corresponding States, which implies that in general the conditions at corresponding states are similar; another expression of the same resemblances is Trouton's rule. Both of these formulations throw some light on the regularities of substances. More insight can be obtained from cases where the law of corresponding states apparently does not hold. We have, for instance, the case of H_2SO_4 and $H_2SO_4 \cdot H_2O$; here the value of ΔB_f for the second compound is about twice that of the first compound, notwithstanding their similarity. These values are hardly an example of the working of the law of corresponding states. However, we can make them meaningful by performing the following calculation:

ΔB_f: H_2SO_4 = 2.7927
ΔB_f: H_2O = 1.7641
one bond = 1
ΔB_f: $H_2SO_4 \cdot H_2O$ = 5.5568 (calc.)
ΔB_f: $H_2SO_4 \cdot H_2O$ = 5.5142 (exp.)

Obviously, the number of bonds broken in going from H_2SO_4 (solid) to H_2SO_4 (liquid) plus the number of bonds broken in going from H_2O (solid) to H_2O (liquid) plus one (for the bond between H_2O and H_2SO_4) is approximately equal to the number of bonds broken in going from solid to liquid $H_2SO_4 \cdot H_2O$. The fact that we had to add one for the bond between the H_2SO_4 and H_2O indicates that in the liquid state $H_2SO_4 \cdot H_2O$ does not exist in appreciable quantity (at least not in the same form as it does in the solid). The calculation also shows that the environment of the H_2O in the liquid state of $H_2SO_4 \cdot H_2O$ is the same as in liquid water. This is extremely interesting information, since it gives us some insight into the nature of the liquid and solid state and also into the nature of the liquid-solid transition.

A similar calculation can be done for HNO_3 and its hydrates:

$$\Delta B_f: \quad HNO_3 \quad = \quad 3.6264$$
$$HNO_3 + H_2O \quad = \quad 5.3905$$
$$HNO_3 + 2H_2O \quad = \quad 7.1546$$
$$HNO_3 + 3H_2O \quad = \quad 8.9187$$
$$\overline{}$$
$$\Delta B_f: \quad HNO_3 \cdot H_2O \quad = \quad 5.9594 \quad (exp.)$$
$$HNO_3 \cdot 3H_2O \quad = \quad 9.1600 \quad (exp.)$$

Here apparently the hydrate is stable in the liquid state; it appears that the trihydrate is more stable than the monohydrate.

Another insight is offered by SO_2 and SO_3 now using ΔB_v

$$\Delta B_v: \quad SO_2 \quad = \quad 7.5922$$
$$\tfrac{1}{2}O_2 \quad = \quad 3.0316$$
$$\overline{\phantom{\tfrac{1}{2}O_2 \quad = \quad 3.0316}}$$
$$SO_3 \quad = \quad 10.6238 \quad (calc.)$$
$$SO_3 \quad = \quad 10.5893 \quad (exp.)$$

Apparently the number of bonds surrounding O in O_2 is the same as the number in SO_3. The question arises, in terms of this formulation, what is the meaning of the Law of Corresponding States? Examining the values of ΔB_v for copper, gold and silver, one sees that the values all lie around 8, the meaning being that the number of nearest neighbors of the metal atom in the liquid just before vaporization is eight. On the other hand, the values of ΔB_v for the noble gases, except helium, are around

six. These values tell us something about the structure of the
liquid at the boiling point. Apparently the free volume of the
liquid increases as we approach the boiling point. Boiling
occurs when the number of nearest neighbors reaches a critical
value, which appears to depend on the structure in the liquid.
Some further light on this concept can be obtained by looking at
the hydrogen halides: For HCl, HBr, and HI the value of ΔB_v
lies around 7; however for HF, the value is around 2. Further,
the values of ΔB_f for the three higher hydrogen halides lie
around 1, while that for HF is about 2. Why? We can offer at
least a partial explanation. As is well known, HF is really a
linear polymer $(HF)_n$, while the other three are not polymers to
any extent. ΔB_v is the number of bonds broken per mole of
unimers vaporized, and we have assumed implicitly in making the
above conclusions that in the vapor state the unimer is the 1-
mer. However, this assumption is not necessarily true, and if
the unimer is not the 1-mer in the vapor, then the reasoning
breaks down. In HF the persistence of the polymer in the solid,
liquid and gas gives rise to abnormal values. If one multiplies
the ΔB_v value of HF by 3.5, one gets values close to those of
HCl, HBr, and HI. We can conclude tentatively that the 1-mer in
the gas phase of HF contains about 3.5 unimers. Similar reasoning leads to some conclusions that can be startling. If one
multiplies the value of ΔB_v of helium by 4 (4 x 1.59 = 6.36),
one gets a value of ΔB_v close to those of the other noble gases.
Does this mean that the 1-mer of He in the gas phase is a tetra-
mer, He_4? A similar conclusion can be obtained from hydrogen,
another substance having apparently an abnormally low ΔB_v value.
If we multiply the ΔB_v value for H_2 by 2, we get 7.11, which is
closer to the more usual values of ΔB_v. The meaning could be
that the 1-mer of hydrogen in the gas phase just above the boiling point is $(H_2)_2$, again a tetrahedral structure. All these
conclusions are based on the idea that in the liquid the unimer
is in a cage of other unimers, the whole group forming the α-mer.
As the α-mer is heated, the value of α decreases and the free
volume increases. As the liquid approaches the boiling point,
it is surrounded by fewer unimers and at each pressure there is
a critical number, from which the 1-mer can escape most easily.

As one can see from Table 6.3, this number varies with the temperature and hence the pressure.

Table 6.3

Values of ΔB_V of mercury as a function of the temperature.
R = 1.98717 cal/mol deg; 0°C = 273.15°K; M.W.(Hg) = 200.59

t°C	H_V cal/g	ΔB_V per mol	
210	71.111	9.905	
250	70.219	9.033	
300	69.116	8.115	
350	68.601	7.408	
357.78	67.844	7.236	N.B.P.
400	66.921	6.690	
450	65.822	6.125	
500	64.713	5.633	
525	64.163	5.410	
535	63.953	5.325	

REFERENCES

Rossini, F.D., Wagman, D.D., Evans, W.H., Levine, S., and Jaffe, I., *Selected Values of Chemical Thermodynamic Properties*, Circular 500, Natl.Bur.Standards (U.S.A.), Series II, 1958.

CHAPTER 7

THE LIQUID STATE AND SURFACE TENSION

Liquids have always been the most poorly understood state of matter, being neither as well ordered as the solid nor as "simple and random" as the gases. Therefore it is not surprising that an immense amount of speculation has appeared in the course of the various attempts to formulate a theory that would describe their nature and properties in a reasonable fashion. We shall not attempt to review this body of work but rather concentrate our attention on a modern description derived from association theory. From this theory we already know that this highly condensed state differs from the solid state, the other highly condensed state, in that, while the liquid state is in 5-symmetry, the solid state is in 6-symmetry. Further, in contrast to gases, liquids consist of extraordinarily large species, which we call α-mers, in equilibrium with 1-mers. It is the 1-mers in the body of the liquid that are responsible for the vapor pressure of the liquid, since they are in equilibrium with the gas phase. However, the liquid has many other properties that need clarification.

EQUATION OF STATE OF THE LIQUID

One of the relationships that needs elucidation is that between the pressure, the temperature, and the volume of a liquid. In the gas phase the original formulation of the law, which is now known as the Ideal Gas Law, gives an approximate relationship between these quantities. Van der Waals, who modified this law, thought that his modification applied both to gases and liquids. While it did indeed represent the behavior of gases better than did the Ideal Gas Law, it was scarcely of any use with liquids, although it did give us some insight into the gas-liquid transition. The equation developed by van der Waals was not, however, entirely satisfactory for practical use with gases. The equation used experimentally was the empirical virial equation, as developed by Kamerlingh Onnes.

In Chapter 3 we derive the association equation of state and show it to be the closed form of the virial equation, which is in the form of an infinite series of which only the first few terms are used. However, the form of the association equation of state that applies to gases needs modification if it is to be used with liquids, which in general have rather high densities. The modification we shall employ comes originally from van der Waals. In his derivation he postulates that the pressure on the gas is not to be represented solely by the external pressure, P_e, but must include a term due to the mutual attraction of the molecules in the gas; i.e., that the pressure term is $(P_e + P_i)$, where P_i is the internal pressure. In classical thermodynamics we equate P_i with $(\partial U/\partial v)_T$ and find that for ideal gases this quantity must be equal to zero. Since van der Waals thought that gases were composed of a large number of identical molecules, his further interpretation was based on this premise and hence need not concern us here.

In a gas, as in a liquid, there is attraction between the elemental units. We have shown that this attractive force gives rise to the different species of particles, and hence the internal pressure must be considered to be the result of this binding force. We shall consequently modify the association equation of state to

$$(P_e + P_i)(v - B) = N_w BRT \qquad (7:1)$$

where P_e is the external pressure, P_i is the internal pressure, N_w, the number of Avogadro units of unimers in the system and the rest of the quantities are as previously defined. In general discussions of this equation N_w is usually taken as equal to one and thus is often omitted.

THE SURFACE LAYER

Liquids act as if they possessed on their surface a layer with special properties; for example, this layer forces a liquid falling through free space to tend to fall in spherical or almost spherical droplets. This behavior has often been formalized into a statement that a liquid tends to assume a shape with the minimum area possible. That this is not a complete statement can be seen experimentally by placing a drop of liquid on a wettable surface. Under such conditions the drop increases its area by spreading on the surface. The formal principle needs the addition of the qualifying phrase: "in the absence of other forces". In the case of the falling drop, the shape is not spherical, because the aerodynamic forces of the air flowing around the drop deform it; in the other case, the spreading film is the result of the force of adhesion of the liquid to the surface. However, this formal statement is only a phenomenological statement of observations and not an explanation of the phenomena. From the viewpoint of association theory the tendency of the liquid, both to assume an almost spherical surface and to spread on a wettable surface, is due to the same cause, namely, the attempt of the unimers in the liquid to become as multiply-bonded as possible, thus losing the maximum amount of energy. It does not matter to the unimers whether they are multiply-bonded to unimers of the same kind or to unimers of another kind. In forming bonds, the unimer loses energy and thus, if it is to bond to different unimers, the energy lost here must be greater than the energy it would lose in bonding to like unimers. In more conventional terms, if it is to spread on a wettable surface, the adhesion energy must be greater than the cohesion energy. In the reverse case the liquid will not spread. No special reason need be called upon to explain these phenomena; the single one, that the unimers try to form a maximum number of multiple bonds with the maximum loss of energy, suffices.

Examining this explanation, we realize that the special properties of the surface are due to the fact that the unimers at the surface are not as multiply-bonded as those in the interior of the liquid. This explanation is not a new one but has long been known. However, there are certain peculiarities in the term "surface"; it is a mathematical abstraction for a two-dimensional film surrounding the liquid. Since unimers are not two-dimensional, the film must have a thickness. The question then becomes: how thick is this film? Various estimates have appeared in the literature. Rusanov estimates that the thickness ranges from 10 to 50Å, although there is no certainty about the exact value. We can arrive at an estimate as follows.

The surface energy of a liquid is usually given as $\gamma(Mv)^{2/3}$, where γ is the surface tension, M is the formula weight of the liquid, and v is the specific volume; (Mv) is the molar volume (formula volume) and $(Mv)^{2/3}$ is the area of the hypothetical two-dimensional surface film. The two thirds power in this expression comes from the fact that, while the volume of a cube is a^3, the surface area is $6a^2$, hence the surface area of one face is the volume to the ⅔ power. But the surface area here is a two-dimensional mathematical surface, which is an untenable abstraction. What is needed is a surface-volume. Hence we write the surface energy as $\gamma(B/d)$, where (B/d) is the surface area (surface film volume) and γ is the surface tension. B is the covolume, which is the actual volume of the particles per formula weight and d is the thickness of the surface film.

Many authors, for example Hildebrand and Scott, have recognized that the internal energy and the surface energy are related, although the exact nature of the relationship is not selfevident. The energy terms are the result of the kinetic energy of the particles, and this does not change at the surface. The only difference between the surface and the interior of the liquid is that in the surface film some of the unimers are not as multiply-bonded as the unimers in the body of the liquid. This difference does not, however, give rise to additional kinetic energy in the system. It thus seems logical to equate the internal energy term to the surface energy term, although an extra term is needed for the work required to form the surface.

$$P_i B = \gamma(B/d) + \text{work term} \qquad (7:2a)$$

$\gamma(B/d)$ is the energy in the surface and $P_i B$ is the energy in the body of the liquid. To evaluate the work term we proceed thus. The work necessary to form the surface is given by the surface tension × the area (the thickness being constant).

$$\text{work term} = \gamma(B/d)$$

Hence Eqn.7:2a becomes

$$P_i B = 2\gamma(B/d) \qquad (7:2b)$$

or

$$d = 2\gamma/P_i \qquad (7:3)$$

d can readily be calculated, if γ and P_i are known. While γ is well known for many substances, P_i, the internal pressure, is a quantity which, although studied extensively, is not known nearly as well as γ. Though tables of P_i are available, all these values have been obtained using some kind of approximation. If we use these values to calculate d, we obtain values as shown in Table 7.1. Although the values of d, the thickness, are remarkably constant, nonetheless their values are much too small from many points of view. In Chapter 8 we shall derive a better value of P_i, which will give a much more reasonable value for d.

Table 7.1

Apparent values* of the thickness coefficient, d, of the surface film at 20°C (see text for comments)   1 cal = 41.2917 cc.atm = 4.184·10^7 ergs

Substance	Internal pressure, P_i			Surface tension	Thickness
	cal/deg[†]	(dynes·10^{-7})/cm^2	Atm	dynes/cm[††]	d Å
Pentane	54.8	229.2	2263	16.05	1.410
Hexane	57.1	238.9	2358	18.40	1.540
Diethyl ether	63.0	263.6	2601	17.10	1.298
Acetophenone	109.3	457.3	4513	39.61	1.732
Propanol	68.8	287.9	2841	23.71	1.648

* see better values next chapter
[†] values from G.Allen, G.Gee and G.J.Wilson
[††] values from J.Jasper

SURFACE TENSION

If we rearrange Eqn.7:1, the equation of state, solving for $P_i B$, we have

$$P_i B = P_e(v - B) + P_i v - BRT \qquad (7:4)$$

omitting the term N_w. Substituting here Eqn.7:2b and rearranging the terms gives

$$\gamma = \frac{P_e(v-B) + P_i v}{2B/d} - \frac{BRT}{2B/d} \qquad (7:5)$$

This equation is in the form

$$\gamma = m - nT \qquad (7:6)$$

with

$$m = [P_e(v-B) + P_i v]/(2B/d) \qquad (7:7)$$

$$n = (RB)/(2B/d) \qquad (7:8)$$

Many determinations of surface tension (see Jasper) have shown that most measurements of surface tension as a function of temperature can be represented in the form

$$\gamma = a - bt \qquad (7:9)$$

where a and b are constants and t is the temperature in degrees Celsius. One can convert this equation readily to the form of Eqn.7:6 by writing

$$\gamma = (a + b \cdot 273.15) - bT = a' - bT \qquad (7:10)$$

If we equate a' to m and b to n, Eqn.7:7 and Eqn.7:3 can be solved for B, and using this value we can solve Eqn.7:8 for B. The derived equations are

$$B = \frac{(P_e + P_i)v}{(a'P_i/\gamma) + P_e} \qquad (7:11)$$

$$B = \frac{bP_i B}{R\gamma} \qquad (7:12)$$

Using the current values of P_i as given by Allen et al. and the values of a and b from Jasper, we have calculated Table 7.2. Another method of solving Eqn.7:11 and 7:12 is available. Equation 7:11 involves P_e and P_i. Along the saturation curve the two values are widely different in value, P_e being much smaller than P_i. If we neglect P_e compared with P_i, then Eqn. 7:11 becomes

$$B = P_i v/(a'P_i/\gamma) = v\gamma/a' \qquad (7:13)$$

Eqn.7:13 gives the same value as Eqn.7:11 to four significant digits. B can now be obtained from Eqn.7:12 or from Eqn.7:1.

Table 7.2

Apparent values* of B and $Ƀ$ at 20°C from the surface tension coefficients, a and b, and the internal pressure, P_i

Substance	$a^{\dagger\dagger}$	$b^{\dagger\dagger}$	$P_e^{\dagger\dagger\dagger}$ (atm)	$\rho^{\dagger}$ (g/cc)	$P_i^{\dagger}$ (atm)	$Ƀ$ cc/FW	B
Pentane	18.25	0.1102	0.553	0.6262	2263	38.24	7.243
Hexane	20.44	0.1022	0.158	0.6595	2358	49.71	7.936
Diethyl ether	18.92	0.0908	0.582	0.7135	2601	40.65	6.841
Acetophenone	41.92	0.1154	**	1.028	4513	63.04	10.102
Propanol	25.26	0.0777	0.0196	0.8036	2841	38.14	4.328
isoButanol	24.53	0.0795	0.0116	0.8070	2725	45.56	5.244

* see better values in next chapter
** values are so low as to be negligible
† from Allen, Gee and Wilson
†† from Jasper
††† from Timmermans

Eqn.7:13 is very interesting in that it shows that $Ƀ$ is not very dependent on P_e or P_i.

The question now presents itself: are the values of B and $Ƀ$ meaningful? That is, is the derivation of the surface tension equation correct and, if it is correct, what is the influence of temperature on the values? A solution to these points can be obtained. In Chapter 6 we had another equation involving B quite independent of the surface tension coefficients. We can connect these two developments and derive the appropriate equation.
From Chapter 6

$$\Delta U = \tfrac{3}{2}R\Delta TB - \int(P_e + P_i)\,d(v - Ƀ) \qquad (7:14)$$

Ww have written here (P_e+P_i) for the P of the equation in the previous chapter in view of our discussion. Classical thermodynamics has the definition

$$H = U + Pv$$

Here too, if we write $(P_e+P_i)(v-Ƀ)$ for Pv, then the definition becomes

$$H = U + (P_e + P_i)(v - Ƀ)$$

or in differential and integrated forms

$$dH = dU + d(P_e + P_i)(v - B)$$
$$\Delta H = \Delta U + \Delta(P_e + P_i)(v - B)$$
(7:15)

Inserting into this equation, the equation of state, Eqn.7:1, we can write

$$\Delta H = \Delta U + R\Delta TB$$
(7:16)

Using Eqn.7:14 and Eqn.7:16, we have

$$\Delta H = \tfrac{5}{2} R \Delta TB - \int_0^T (P_e + P_i)\, d(v - B)$$

We take the lower limit of the integration to be 0°K, since we are interested in the bonds broken from absolute zero. At 0°K the value of H is H_0, while the value of BRT is zero. Hence

$$\Delta H = H - H_0 = \tfrac{5}{2} BRT - \int_0^T (P_e + P_i)\, d(v - B) \quad (7:17)$$

$\Delta H = H - H_0$ is known in the literature for a number of substances.*
To calculate the value of B from this equation, we must also know the value of the integral on the right-hand side of the equation. As a first approximation, we shall consider it to be zero. This is a rather presumptious assumption, since P_i is quite large; however, as usually presented, the term is Pdv and in the liquid state P is taken to be equal to the external pressure. Since the density is high, the term is small and hence is neglected. Later we shall explore the impact of this approximation.

Especially good values of $H-H_0$ for a few substances are known from the work done in the National Physical Laboratory in Teddington, England. Using these values, we have calculated B from $H-H_0$, setting the integral equal to zero. These values are given in Table 7.3, together with the values of B calculated from the surface tension and P_i. As we can see, these values are approximately equal. However, we know that the value for $H-H_0$ contains a serious approximation. The value determined by

---

* In general, $H-H_0$ is a quantity difficult to determine, as its rate of variation near transition points is high. Near constant-temperature phase changes the change in $H-H_0$ becomes especially great, and hence $H-H_0$ increases rapidly as the transition temperature is approached.

the other method contains an equivalent approximation in the determination of P_i. Both approximations are similar but enter in a slightly different manner; hence the closeness in value of B determined by both methods. Let us then examine these approximations.

Table 7.3

Values of B calculated from $H-H_0$ and from surface tension and P_i at 20°C

Substance	B from $H-H_0$	B from γ & P_i
Propyl alcohol	4.672	4.328
iso-Butyl alcohol	5.385	5.244
Diethyl ether	6.175	6.841

The internal pressure, P_i, is a quantity which has long been known in thermodynamics but whose magnitude is difficult to determine. Numerous methods have been devised to evaluate it, but all of them have included some approximation or another. The values obtained do not agree and many investigators feel that, in general, the values obtained are too large. The method currently in use, by which the values in Table 7.2 are calculated, uses experimental thermal expansivity and the thermal compressibility. The necessary formulae are derived thus: we start with the classical thermodynamic equation of state

$$\left(\frac{\partial U}{\partial v}\right)_T = T\left(\frac{\partial p}{\partial T}\right)_v - p \qquad (7:18)$$

where p is the external pressure. The differential $(\partial U/\partial v)_T$ has long been known as the internal pressure. Thus, if we make this substitution, and realizing that p, the external pressure, is small for liquids, we get the equation

$$P_i \simeq T\left(\frac{\partial p}{\partial T}\right)_v \qquad (7:19)$$

The differential in this equation is determined directly only very infrequently, it being a rather difficult experiment. Generally, it is evaluated using the identity that

$$\left(\frac{\partial p}{\partial T}\right)_v = -\frac{(\partial v/\partial T)_p}{(\partial v/\partial p)_T}$$

Since $\quad \alpha = \frac{1}{v_0}\left(\frac{\partial v}{\partial T}\right)_p \quad$ and $\quad \beta_T = -\frac{1}{v_0}\left(\frac{\partial v}{\partial p}\right)_T \quad$ (7:20)

$$P_i \simeq T\frac{\alpha}{\beta_T}$$

Again β_T is commonly not obtained directly but rather from the velocity of sound; i.e.,

$$\beta_s = 1/\rho v^2 \quad \text{and} \quad \beta_T = \beta_s + (T\alpha^2)/(\rho C_p) \quad (7:21)$$

where β_s is the adiabatic compressibility, ρ is the density, and C_p is the molar specific heat.

This is a complicated web of equations. The nature of the approximation can be seen much more directly if we examine the $H-H_0$ approximation. This will enable us to examine what effect the various approximations have on the value of B. Solving Eqn.7:17 for B, we have

$$B = \frac{(H-H_0) + \int(P_e+P_i)d(v-B)}{5/2\,RT} \quad (7:22)$$

Solving Eqn.7:12 for P_i gives

$$P_i = (BR\gamma)/(bB) \quad (7:23)$$

which is substituted into Eqn.7:11; after using Eqn.7:10 and solving, we have

$$B^2 + B \cdot \theta - \phi = 0 \quad (7:24)$$

From this

$$B = -\tfrac{1}{2}\theta + \tfrac{1}{2}\sqrt{\theta^2+4\phi}$$

with $\quad \phi = (RB\gamma v)/(P_e b) \quad (7:25)$

$$\theta = \frac{BRa'}{P_e b} - v$$

Table 7.4

Values of B, $\bar{B}$ and P_i from $H-H_0$ [§] and the surface tension coefficients, a and b, using $\int(P_e+P_i)\,d(v-\bar{B}) \simeq (P_e+P_i)(v-\bar{B})$

t°C	$H-H_0$ joules	P_e atm	v cc/FW	B	$\bar{B}$ cc/FW	P_i atm
Diethyl ether ($a = 18.92$; $b = 0.0908$)						
0	34306.96[†]	0.244*	100.68**	4.3161	43.573	1693.6
20	37627.40	0.582	103.89	4.4108	40.649	1677.2
25	38513.27	0.707	104.73	4.4390	39.892	1674.3
34.43	40679.1	1.00	106.43	4.5448	38.459	1686.7
1,1,1 Trichlorethane ($a = 28.28$; $b = 0.1242$)						
0	31640[†]	0.0504*	97.33**	3.9805	44.248	1680.8
20	34499	0.138	99.74	4.0442	41.362	1666.4
25	35220	0.172	100.36	4.0594	40.619	1662.2
iso-Butyl alcohol ($a = 24.53$; $b = 0.0795$)						
16.85	32250[†]	0.00963*	92.12**	3.8215	46.194	1980.2
20	32812	0.0116	92.41	3.8464	45.841	1986.7
25	33710	0.0145	92.88	3.8854	45.277	1996.6
36.85	35910	0.0317	94.02	3.9807	43.916	2020.9
46.85	37870	0.0584	95.00	4.0668	42.742	2043.2
n-Propyl alcohol ($a = 25.26$; $b = 0.0777$)						
0	25750[†]	0.00453*	73.35**	3.2395	39.860	2168.0
20	28469	0.0191	74.79	3.3372	38.143	2190.4
25	29180	0.0265	75.15	3.3632	37.697	2196.9
36.85	30920	0.0550	76.06	3.4276	36.646	2212.2
46.85	32460	0.0967	76.89	3.4858	35.762	2225.5
Nitrogen liquid ($a = -35.45$; $b = 0.2265$)						
tp -210.01	2698.68[††]	0.1235[††]	32.27[††]	1.4688	14.805	435.56
-203.15	3082.77	0.3808	33.36	1.5134	13.349	433.93
bp -195.80	3482.76	1.00	34.68	1.5473	11.701	426.77
Oxygen liquid ($a = -33.72$; $b = 0.2561$)						
tp -218.80	2578.18[††]	0.001474[††]	24.84[††]	1.6301	15.296	761.95
-203.15	3411.63	0.06182	26.10	1.6748	13.188	744.91
-193.15	3946.35	0.2974	27.05	1.6952	11.758	727.32
bp -182.97	4493.62	1.00	28.19	1.7123	10.233	704.95

Table 7.4 (continued)

t°C	$H-H_0$ joules	P_e atm	v cc/FW	B	Þ cc/FW	P_i atm
Argon liquid ($a = -33.82$; $b = 0.2493$)						
tp −189.34	2874.83[††]	0.6801[††]	28.231[††]	1.1788	11.034	470.69
−188.15	2976.29	0.7787	28.391	1.1831	10.852	469.68
bp −185.85	3027.96	1.00	28.699	1.1919	10.492	467.97

tp = triple point; bp = boiling point

[§] All variables were inserted with the maximum number of digits possible. Certain data were given at odd intervals. Such data were interpolated using 5-or 7-point Lagrangian interpolation.

[†] See listing under National Physical Laboratory, Teddington, England
[*] Values of P_e interpolated from Timmermans
[**] Data calculated from Allen et al.
[††] Nitrogen, oxygen and argon data from Hultgren, Desai et al.

In doing the calculation, the value of B is first obtained from Eqn.7:22 by setting the value of the integral, $\int (P_e + P_i) d(v - b)$, equal to zero. Then a solution is obtained for Þ from Eqn.7:24. The solution for P_i comes from Eqn.7:23 or Eqn.7:1. One could now use the values of P_e, P_i, and Þ to obtain a second approximation for B if one knew how to evaluate the integral. Unfortunately, we do not. We shall therefore use the following stratagem. P_i (as we shall see later) is only a slowly varying quantity; hence we shall consider the integration to proceed as follows:

$$\int (P_e + P_i) d(v - \text{Þ}) \approx (P_e + P_i) \Delta(v - \text{Þ}) \approx (P_e + P_i)(v - \text{Þ}) \quad (7:26)$$

since the free volume, $(v-\text{Þ})$, is probably equal to zero at 0°K. This correction factor is probably rather too small, but making this approximation enables us to evaluate the expression. This procedure gives us a new value for B, from which in turn we obtain new values of Þ and P_i. This iteration is repeated and converges; after about 25 steps it gives agreement in B to 10^{-6}. The values for this calculation are shown in Table 7.4.

Comparing the values of P_i in Table 7.2 and Table 7.4, we can see that the approximation was extensive. If, instead of using $p = (P_e+P_i)$, we had ignored the existence of the internal pressure and taken the total pressure to be the external pressure, the seriousness of the approximation would have been much less. One could go into further discussion here, but it would be futile, since in the next chapter an exact method of obtaining P_i, free of approximations, will be described. This will show that even the approach used here underestimates the value of the integral.

IMPORTANT EQUATIONS

$$(P_e + P_i)(v - B) = N_w BRT \tag{7:1}$$

$$d = \frac{2\gamma}{P_i} \tag{7:3}$$

$$\gamma = a - bt = a' - bT \tag{7:9}\&(7:10)$$

$$B = \frac{(P_e + P_i)v}{(a'P_i/\gamma)+P_e} \tag{7:11}$$

$$B = \frac{bP_i B}{R\gamma} \tag{7:12}$$

$$\Delta H = H - H_0 = \tfrac{5}{2} BRT - \int_0^T (P_e + P_i)\, d(v - B) \tag{7:17}$$

REFERENCES

Allen, G., Gee, G. and Wilson, G.J., *Internal pressure and cohesive energy: densities of simple liquids*, Polymer (London), 1 (1960) 456-466.

Boks, J.D.A. and Kamerlingh Onnes, Onnes Communications, No.170a, Leiden, 1924.

Hildebrand, J.H. and Scott, R.L., *The Solubility of Nonelectrolytes*, 3rd edn., Reinhold, New York, 1950.

Jasper, J.J., *The surface tension of pure liquid compounds*, J. Phys. Chem. Ref. Data, 1 (1972) 841-1010.

Hultgren, R., Desai, P.D., Hawkins, D.T., Gleiser, M., Kelley, K.K. and Wagman, D.D., *Selected Values of the Thermodynamic Properties of the Elements*, American Society for Metals, Metals Park, Ohio, 1973.

Rusanov, A.I., *Recent investigations on the thickness of surface layers*,
    Prog. Surf. Membr. Sci., 4 (1971) 57-114.

Timmermans, J., *Physical Chemical Constants of Pure Organic Compounds*,
    Elsevier, Amsterdam, Vol. 1, 1950, Vol. 2, 1965.

Van der Waals, J.D., *De continuiteit v. d. vloeibaaren en gasvormigen
    toestand*, (Diss.) Leiden, 1873 (trans. into German by F. Roth,
    Leipzig, 1881).

Values from the National Physical Laboratory, Teddington, England,
    1,1,1 trichlorethane,
        Anderson, R.J., Counsell, J.F., Lee, D.A. and Martin, J.F.,
        J. Chem. Soc. Faraday, 69 (a) (1973) 1721-1726.

    iso-butyl alcohol and n-propyl alcohol,
        Counsell, J.F., Lees, E.B. and Martin, J.F.,
        J. Chem. Soc., (A) (1968) 1819-1823.

    diethyl ether
        Martin, J.F., Counsell, J.F. and Lees, E.B.,
        J. Chem. Soc., (A) 21 (1971) 313-316.

CHAPTER 8

THE TAIT-TAMMANN EQUATION

In the late 1800's P.G.Tait, a noted English scientist, investigated the compressibility of water and developed an empirical equation to represent its behavior. The equation is a notable one; for, although the liquid state was not understood and calculations about it and its properties were very approximate, this equation shone like a beacon in its exactness and offered hope that the liquid state could be made understandable. Further, the equation applied not only to water, a notoriously bad actor, inasmuch as its properties are peculiar, but also to a whole variety of other liquids. It was shown by Carl and Wohl to apply to methyl, ethyl, propyl, isopropyl and allyl alcohols, acetone, ethyl ether, ethylchloride, bromide and iodide, carbon disulfide, phosphorus trichloride; by Gibson and coworkers to benzene, chloro-, bromo- and nitro-benzene, aniline, ethylene glycol and its water solutions, and to concentrated solutions of sodium chloride and sodium bromide; by Ginell to liquid helium I and II, etc. In fact, Hirschfelder, Curtis and Bird in their monumental book on liquids state that "studies of the properties of a large number of liquids have shown that this equation gives almost perfect agreement with experimental observations". Moreover, the equation not only holds for liquids but it also applies to solids like the alkali metals as shown by Ginell and Quigley, and to glasses as given by Brown and Ginell.

However, this equation, being empirical and hence not having theoretical underpinnings, was vulnerable to attack, and indeed it does have its detractors. Hayward, for example, claims that, since liquids in general are highly incompressible, any linear type of equation might be fitted to the data, and Macdonald does in fact fit other linear equations to the data. Some of the equations are similar to the Tait-Tammann equation and others differ. Sometimes the fit was better than the fit of the Tait-Tammann* equation, although the term "better" should be put in quotes, for the compressibility is small in any case, the data contain their usual assortment of experimental errors, and the criterion for "better" is statistical. It is always a difficult task to choose between empirical equations on the basis of their fit to experimental data. The best criterion for choosing an equation is that the equation chosen relates to the rest of theoretical science. While it is often expedient to use an empirical equation for interpolation in a set of data or even for small extrapolations, scientists feel more comfortable with equations that have meaning in the larger sense and that give them insight into the phenomena.

In the next section we shall derive the Tait-Tammann equation and relate it to the rest of the theory.

THE DERIVATION

We start with the equation of state as developed in Chapter 7 and divide by v

$$(P_e + P_i)(v - \mathrm{B}) = N_w RTB$$

$$(P_e + P_i)(1 - \mathrm{B}/v) = N_w RTB \quad \text{with } \mathrm{B} = B/v \quad (8:1)$$

Differentiating, keeping T constant and rearranging, we have

$$\left(\frac{\partial}{\partial v}(P_e + P_i)\right)_T = \frac{N_w RT \left(\frac{\partial \mathrm{B}}{\partial v}\right)_T - (P_e + P_i)\left(\frac{\partial}{\partial v}(1 - \mathrm{B}/v)\right)_T}{(1 - \mathrm{B}/v)} \quad (8:2)$$

Carrying out the operation on the left-hand side and rearranging, we have

$$\left(\frac{\partial}{\partial v} P_e\right)_T = \frac{N_w RT \left(\frac{\partial \mathrm{B}}{\partial v}\right)_T - (P_e + P_i)\left(\frac{\partial}{\partial v}(1-\mathrm{B}/v)\right)_T - (1-\mathrm{B}/v)\left(\frac{\partial}{\partial v} P_i\right)_T}{(1 - \mathrm{B}/v)} \quad (8:3)$$

---

* As pointed out by Hayward, the original Tait equation is an equation in integrated form. The differential equation now often labeled the Tait equation was apparently first used by Tammann and is given in one of his books. We have here labeled this differential equation the Tait-Tammann equation for clarity.

Inverting the equation and dividing top and bottom of the right-hand side by $\frac{\partial}{\partial v}(1-E/v)$ and changing sign gives

$$-\left(\frac{\partial v}{\partial P_e}\right)_T = \frac{\frac{(1-E/v)}{\frac{\partial(1-E/v)}{\partial v}}}{\left[\frac{-N_w RT\left(\frac{\partial B}{\partial v}\right)}{\frac{\partial(1-E/v)}{\partial v}} + \frac{(1-E/v)\left(\frac{\partial P_i}{\partial v}\right)}{\frac{\partial(1-E/v)}{\partial v}} + P_i\right] + P_e} \qquad (8:4)$$

Calling

$$\frac{(1-E/v)}{\partial(1-E/v)/\partial v} = J \qquad (8:5a)$$

and

$$\frac{-N_w RT(\partial B/\partial v)+(1-E/v)(\partial P_i/\partial v)+P_i[\partial(1-E/v)/\partial v]}{\partial(1-E/v)/\partial v} = L \qquad (8:5b)$$

We have

$$-\left(\frac{\partial v}{\partial P_e}\right)_T = \frac{J}{L+P_e} \qquad (8:6)$$

This equation is in exactly the form of the Tait equation as modified by Tammann. In the Tait-Tammann equation J and L have been found to be experimentally constant with a change in pressure, although they may vary with the temperature. Let us then examine the combination of factors that comprise J and L and see whether these expressions, Eqns. 8:5a and 8:5b, can be simplified so as to obtain useful relationships from them.

THE CONSTANT L

The simplification of L is straightforward, although a little complicated. Equation 8:5b for L contains a term for $\partial B/\partial v$. Let us first simplify this term. From Eqn. 7:12 we have

$$\left(\frac{\partial B}{\partial v}\right)_T = \frac{\partial}{\partial v}\left(\frac{bP_i(E/v)}{N_w R\gamma}\right)_T$$

$$\left(\frac{\partial B}{\partial v}\right)_T = \frac{b}{N_w R\gamma}\left(\frac{E\partial P_i}{v \partial v} + P_i \frac{\partial(E/v)}{\partial v}\right)_T \qquad (8:7)$$

Substituting this equation into Eqn. 8:5b and rearranging, we have

$$\frac{\partial P_i}{\partial v}\left[-\frac{bT}{\gamma}\left(\frac{E/v}{\frac{\partial(1-E/v)}{\partial v}}\right)+\left(\frac{(1-E/v)}{\frac{\partial(1-E/v)}{\partial v}}\right)\right] + P_i(1+bT/\gamma) = L \qquad (8:8)$$

This equation can be arranged to be in the form $dx/dy+Mx = Q$. This of course is a linear differential equation whose solution is known to be

$$x = \exp(-\int Mdv)[\int(Q\exp\int Mdv)dv+\text{const}] \qquad (8:9)$$

In Eqn.8.10 after some rearranging of the terms, we have

$$M = \frac{\left[1+\frac{bT}{\gamma}\right]\left[\frac{\partial}{\partial v}(1-E/v)\right]}{1-\left[1+\frac{bT}{\gamma}\right]E/v} \qquad (8:10)$$

$$Q = \frac{L\left[\frac{\partial}{\partial v}(1-E/v)\right]}{1-\left[1+\frac{bT}{\gamma}\right]E/v} \qquad (8:11)$$

Looking at the integral in Eqn.8:9

$$\int Mdv = \int \frac{(1+bT/\gamma)d(1-E/v)}{1-(1+bT/\gamma)E/v} = \ln[1-(1+bT/\gamma)E/v] \qquad (8:12)$$

we have with the proper substitution and integration from Eqns.8:10 and 8:11

$$P_i = \frac{1}{1-(1+bT/\gamma)(E/v)}[L(1-E/v)+\text{const}] \qquad (8:13)$$

From Eqn.7:11, we have

$$\frac{E}{v} = \frac{P_e+P_i}{(a'P_i/\gamma)+P_e} \qquad (8:14)$$

When $P_e = 0$, then

$$E/v = \gamma/a' \qquad (8:15)$$

Using this equation and the equation

$$\gamma = a' - bT \qquad (8:16)$$

Eqn.8:13 gives

$$\text{constant} = L\left(\frac{\gamma}{a'} - 1\right) \qquad (8:17)$$

Hence Eqn.8:13 then becomes

$$P_i = \frac{L(\gamma/a' - E/v)}{(1 - a'E/v\gamma)} = \frac{\gamma}{a'} L \tag{8:18}$$

L is an empirical constant, therefore

$$\frac{a'}{\gamma} P_i = L$$

We must now see how these three factors behave. From Eqn. 8:14 and Eqn. 8:18 we have

$$\frac{E}{v} = \frac{P_e + P_i}{L + P_e} \tag{8:19}$$

Further then

$$1 - \frac{E}{v} = 1 - \frac{P_e + P_i}{L + P_e} = \frac{L - P_i}{L + P_e} \tag{8:20}$$

This determination of the boundary condition is not entirely exact. If one rewrites Eqn. 8:14 as

$$\frac{E}{v} = \frac{1 + (P_e/P_i)}{(a'/\gamma) + (P_e/P_i)} \tag{8:14a}$$

one sees that E/v is equal to γ/a' when P_e/P_i is equal to zero. When this quantity is small, approximating zero, then the boundary condition is real. But the lowest pressure possible in a condensed system is the saturation pressure, and this is not zero, although at many temperatures it is close to zero. Hence under such conditions the boundary condition is valid and leads to no ambiguity. In this connection see also Chapter 10.

THE CONSTANT J

Equation 8:5a can readily be simplified. Rewriting it, we have

$$\frac{\partial(1 - E/v)}{\partial v} = \frac{(1 - E/v)}{J}$$

This equation can be integrated after separating the variables to give

$$(1 - E/v) = A e^{v/J} \tag{8:21}$$

where A is a constant of integration. Since the integration is at constant temperature, A is constant with pressure and its variation with temperature depends on other factors.

DETERMINATION OF J AND L

The determination of J and L is not too difficult. If one inverts the Tait-Tammann equation, one has

$$-\frac{dP_e}{dv} = \frac{L}{J} + \frac{P_e}{J} \tag{8:22}$$

A plot of dP_e/dv versus P_e yields a straight line with a slope $1/J$ and an intercept L/J, thus making both J and L readily available. However simple in principle the calculation appears to be, it offers some difficulties in execution.

The isothermal compressibility is not a quantity usually determined. It is generally obtained indirectly from the velocity of sound with some approximations. We have found that it is best to derive this parameter from the isothermal values of P_e as a function of v. These values are found much more often in the literature. This leaves us with the necessity of obtaining the derivative numerically. We can do this in a primitive fashion by drawing the curve of P_e vs v and determining the tangent graphically at various points. This procedure is both time-consuming, tedious and imprecise, although in the early stages we did use this procedure and derived fair results. A better method is to proceed as follows. The P-v values are converted into a set of points with equally spaced intervals. If the data is reasonably good this can readily be done by numerical interpolation. We have used Lagrangian interpolation, which gives excellent results. (The Fortran program for this procedure is to be found in the Appendix.) The derivative at each of the interpolated points is then found by another program for numerical differentiation. We have used Lagrangian differentiation (see Appendix). Used together, these programs give us a smooth set of values, from which the values of J and L can be determined by a least squares calculation. Using this method, numerous values of J and L have been determined. However, when we evaluate J and L for a substance at various temperatures, the variation with temperature is often not very consistent. This is the result of the statistical nature of the best of experimental data. Further, when we use the integrated Tait-Tammann equation, we must have the value of the constant H (see Eqn.8:25). When

H is evaluated at a given temperature at each pressure in the set, the results are different and hence the resulting H's are averaged. As a result of this type of calculation the errors in J, L and H are different. We can remedy this by using a statistical method outlined by Deming (see Appendix), which uses the values of J, L and H thus derived as a first approximation, carries out a sophisticated least squares calculation, and thus minimizes the error in all three parameters simultaneously. We compute this Deming correction repeatedly using the results of the first calculation to give us a second approximation, etc. If the data is reasonably consistent, the process converges after a number of iterations. We have our tolerances such that we consider that we have convergence when successive approximations agree within 0.0001% for each parameter. This, generally, takes about 6-8 iterations. If one sets the limits of convergence to be lower, then convergence is very often not reached, because the original round-off error of the data starts to play a role. In many sets of data the iteration by the Deming method either does not converge or it converges only after a very large number of iterations at values that are entirely unreasonable when viewed in the context of the adjoining values, although they are valid from a strictly mathematical point of view. Such values are discarded and not reported here, except in rare instances, which are noted in the tables. In Table 8.1 are to be seen the effects of the Deming iteration. The adjustment in H is the greatest, because in the original approximation, H had the greatest burden of error, being the last quantity to be determined. It can be seen from Table 8.1 that the adjustment in H is opposite in sign to the adjustment in J and L and that they determine its magnitude. The values that result from this adjustment are greatly improved, the trends are in the main consistent in direction, although some small anomalies still exist, probably due to systematic errors of some sort. Values of J, L and H are given in Table 8.2. Figures 8.1 and 8.2 show graphically the values of J and L for some substances whose values cover a large range of temperatures. It is often stated in the literature that $C = J/v_0$ is constant with temperature. Looking at Fig. 8.1, one sees that this might be considered to be true for water. However, it hardly seems likely that the trend in

the values for diethyl ether is due entirely to a systematic
error in the data. In any case, we at no time assume that C
(or J) is constant with temperature. The behavior of L around
the critical point is considered in Chapter 10.

Table 8.1

Changes due to the Deming iteration on the parameters J, L and H

Diethyl ether: Amagat data

Temp. °C	$\Delta J/J$ %	$\Delta L/L$ %	$\Delta H/H$ %
0.0	+ 2.3	+ 6.3	− 21.9
10.80	+ 2.7	+ 7.5	− 26.7
20.20	− 2.2	− 6.5	+ 16.6
30.35	+ 0.65	+ 1.7	− 5.7
40.45	+ 3.8	+ 13.3	− 37.2
49.95	+ 1.2	+ 4.9	− 9.5

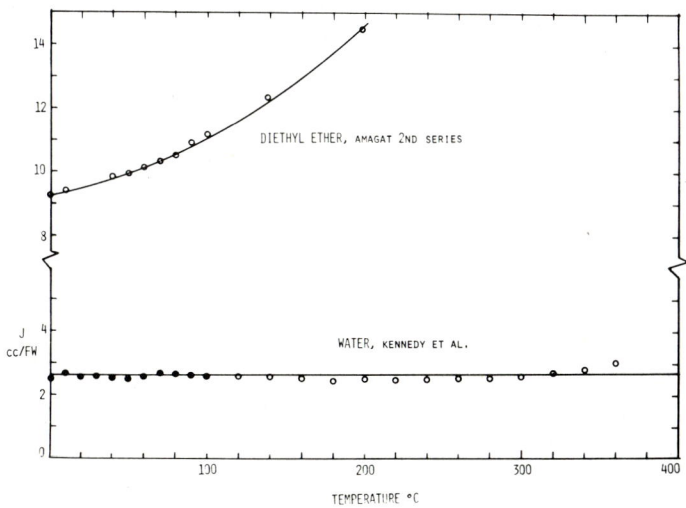

Fig.8.1. J, cc/FW vs. temperature, °C.
● Kennedy, Knight & Holser data, ○ Holser & Kennedy data, and
diethyl ether, Amagat data. The Amagat data for water agree with
the Kennedy et al. data, although they show a little greater spread.

Table 8.2

Experimental Tait-Tammann coefficients and derived quantities for various substances

PROD = <u>P</u>ressure <u>R</u>ange of <u>O</u>riginal <u>D</u>ata

$\gamma = a' - bT$

Experimental quantities				Derived quantities			
t°C	J cc/mole	L atm	H atm $\times 10^{-4}$	P_i atm	$A \times 10^{+4}$	d Å	σ of v Eqn.8:25
Methyl alcohol	Amagat data	(a' = 45.164	b = 0.07730)				
0.0	3.99044	1051.266	2117.35	599.8	0.2321	8.480	0.0051
9.2	4.08886	1015.775	1777.25	524.9	0.2762	8.776	0.0042
23.85	4.15616	930.526	1626.93	457.5	0.2907	9.580	0.0054
30.05	4.27934	877.799	1353.06	422.3	0.3367	10.156	0.0084
	PROD = 1 - 3000 atm						
Ethyl alcohol	Amagat data - first series			(a' = 46.776	b = 0.0832)		
0.0	5.55638	1037.346	3031.92	533.4	0.1662	8.900	0.0055
9.60	5.62335	982.523	2811.32	488.4	0.1758	9.397	0.0126
19.80	5.66125	915.101	2732.58	438.3	0.1745	10.089	0.0113
30.65	5.73407	853.212	2523.86	392.2	0.1827	10.821	0.0141
40.35	5.76478	782.152	2470.12	346.0	0.1766	11.804	0.0201
	PROD = 1 - 3000 atm						
Ethyl alcohol	Amagat data - second series			(a' = 46.776	b = 0.0832)		
0.0	5.21688	941.944	5381.89	484.3	0.0850	9.801	0.0029
20.0	5.39073	842.143	4236.45	403.0	0.1037	10.964	0.0058
40.0	5.42384	719.059	4306.28	318.5	0.0930	12.840	0.0039
60.0	5.55241	618.995	3723.04	252.2	0.0985	14.916	0.0053
80.0	5.68228	521.862	3239.17	194.1	0.1012	17.692	0.0025
100.0	6.03149	455.655	2022.45	153.2	0.1495	20.263	0.0029
198.0	7.53128	59.644	461.81	9.7	0.1082	154.802	0.0145
	PROD = 1 - 1000 atm						

Table 8.2 (continued)

Experimental quantities				Derived quantities			
t°C	J cc/mole	L atm	H atm $\times 10^{-4}$	P_i atm	$A \times 10^{+4}$	d Å	σ of v Eqn.8:25

n-Propyl alcohol   Amagat data   (a' = 46.484 b = 0.0777)

0.0	7.14666	1159.676	4054.15	630.2	0.1306	7.912	0.0105
9.75	7.17616	1078.051	3989.74	568.3	0.1278	8.511	0.0124
25.25	7.33929	998.680	3418.92	500.5	0.1457	9.187	0.0248
39.75	7.30742	866.346	3672.55	413.2	0.1234	10.591	0.0263

PROD = 1 - 2700 atm

Allyl alcohol   Amagat data   (a' = 52.168 b = 0.09020)

| -41.40 | 6.64508 | 1217.853 | 3151.13 | 729.9 | 0.1549 | 8.455 | 0.0121 |
| 35.45 | 6.55151 | 963.672 | 3835.58 | 449.5 | 0.1341 | 10.685 | 0.0076 |

PROD = 1 - 2700 atm

Acetone   Amagat data   (a' = 56.853 b = 0.1120)

| 0.0 | 7.21681 | 1030.121 | 2057.13 | 475.81 | 0.2694 | 10.894 | 0.0163 |
| 35.05 | 7.43200 | 727.992 | 1789.07 | 285.99 | 0.2471 | 15.415 | 0.0175 |

PROD = 1 - 2800 atm

Ethyl bromide   Amagat data   (a' = 58.178 b = 0.1159)

| 10.1 | 7.49189 | 885.450 | 2139.95 | 385.8 | 0.2335 | 12.969 | 0.0089 |
| 42.25 | 7.98424 | 690.856 | 1391.47 | 256.8 | 0.3120 | 16.622 | 0.0350 |

PROD = 1 - 2400 atm

Ethyl iodide   Amagat data   (a' = 66.797 b = 0.1286)

| 10.6 | 8.60051 | 1209.445 | 1579.76 | 548.7 | 0.4182 | 10.901 | 0.0128 |
| 42.55 | 8.44217 | 865.205 | 1979.20 | 339.3 | 0.2657 | 15.239 | 0.0151 |

PROD = 1 - 3000 atm

Phosphorus trichloride   Amagat data   (a' = 65.721 b = 0.1266)

| 10.1 | 9.00726 | 1180.088 | 2131.48 | 536.2 | 0.3021 | 10.993 | 0.0149 |
| 42.1 | 9.17351 | 894.630 | 1973.36 | 351.3 | 0.2753 | 14.500 | 0.0283 |

PROD = 1 - 3000 atm

Table 8.2 (continued)

Experimental quantities				Derived quantities			
t°C	J cc/mole	L atm	H atm $\times 10^{-4}$	P_i atm	$A \times 10^{+4}$	d Å	σ of v Eqn.8:25

Carbon disulfide   Amagat data   ($a' = 75.825$ $b = 0.1484$)

t°C	J	L	H	P_i	A	d	σ
0.0	6.28877	1381.032	1610.31	642.75	0.4585	10.837	0.0065
9.15	6.28523	1285.723	1654.55	575.37	0.4293	11.641	0.0094
19.35	6.47634	1251.227	1345.77	534.95	0.5322	11.962	0.0038
29.95	6.54621	1161.122	1269.19	472.34	0.5427	12.890	0.0045
41.25	6.61432	1058.618	1198.33	407.23	0.5436	14.138	0.0059
49.15	7.14126	1181.148	705.86	436.10	1.0555	12.671	0.0029

PROD = 1 - 3000 atm

Diethyl ether   Amagat data - first series   ($a' = 43.722$ $b = 0.0908$)

t°C	J	L	H	P_i	A	d	σ
0.0	9.55703	675.856	2520.97	292.5	0.1521	12.769	0.0258
10.80	9.65743	595.621	2378.35	244.4	0.1477	14.489	0.0197
20.20	9.74046	529.583	2271.43	207.0	0.1420	16.296	0.0174
30.35	9.95284	477.621	1948.28	176.6	0.1545	18.069	0.0262
40.45	10.34618	465.666	1463.61	162.4	0.2072	18.533	0.0154
49.95	10.59823	436.348	1245.53	143.6	0.2351	19.778	0.0193

Prod = 1 - 3000 atm

Diethyl ether   Amagat data - second series   ($a' = 43.722$ $b = 0.0908$)

t°C	J	L	H	P_i	A	d	σ
0.0	9.28300	623.439	3196.21	269.8	0.1107	13.843	0.0071
10.0	9.40265	561.415	2959.20	231.3	0.1116	15.372	0.0061
40.0	9.86476	399.191	2152.03	139.6	0.1206	21.619	0.0087
50.0	9.94257	342.858	2088.57	112.8	0.1102	25.171	0.0074
60.0	10.13195	300.306	1838.32	92.53	0.1130	28.738	0.0078
70.0	10.32896	259.030	1613.33	74.44	0.1144	33.317	0.0077
80.0	10.50128	219.756	1453.78	58.59	0.1109	39.271	0.0099
90.0	10.91387	194.110	1078.03	47.72	0.1358	44.460	0.0239
100.0	11.17072	163.102	921.95	36.71	0.1371	52.912	0.0154
138.0	12.31761	62.908	479.362	9.19	0.1121	137.186	0.0143
198.0*	14.50341	-55.573	175.138	-1.20	-0.3105	-155.291	0.0734

PROD = 1 - 1000 atm

* above critical temperature

Table 8.2 (continued)

Experimental quantities				Derived quantities			
t°C	J cc/mole	L atm	H atm $\times 10^{-4}$	P_i atm	A $\times 10^{+4}$	d Å	σ of v Eqn.8:25
Water	Amagat data - first series			(a' = 116.174	b = 0.1477)		
0.0	2.23462	2336.31	763.691	1525.0	1.6821	9.815	0.0015
2.10	2.25324	2408.89	715.472	1565.9	1.8132	9.519	0.0017
4.35	2.28161	2483.83	667.670	1607.5	1.9122	9.232	0.0014
6.85	2.31380	2565.09	618.079	1652.0	3.1048	8.940	0.0012
10.10	2.33761	2638.80	588.413	1688.5	2.4514	8.690	0.0011
	PROD = 1 - 3000 atm						
Water	Amagat data - second series			(a' = 116.174	b = 0.1477)		
0.0	2.49547	2666.707	368.091	1740.6	2.5159	8.599	0.000243
5.0	2.81653	3163.919	191.330	2045.1	5.8478	7.248	0.000764
10.0	2.51932	2866.755	369.849	1834.8	2.7903	7.999	0.000213
15.0	2.50898	2918.077	389.777	1849.1	2.7426	7.858	0.000416
20.0	2.58703	3074.803	332.541	1928.8	3.4461	7.458	0.000501
30.0	2.68252	3279.884	281.345	2015.8	4.4931	6.991	0.000613
40.0	2.65729	3282.343	307.211	1975.5	4.2537	6.986	0.000554
50.0	2.59800	3200.035	360.614	1885.3	3.6457	7.166	0.000482
*							
80.0	2.19431	2488.569	1174.842	1371.2	0.9510	9.215	0.000293
100.0	2.38888	2562.263	676.409	1346.7	1.7971	8.950	0.000863
198.0	2.60587	1423.226	433.899	570.7	1.9648	16.112	0.00123
	PROD = 1 - 1000 atm		Average $J\pm\sigma$ = 2.550±0.163				

* 60°, 70°, and 90° values rejected for non convergence

Table 8.2 (continued)

Experimental quantities				Derived quantities			
t°C	J cc/mole	L bars	H bars $\times 10^{-4}$	P_i bars	$A \times 10^{+4}$	d Å	σ of v Eqn.8:25
Water	Kennedy, Knight and Holser data			(a' = 116.174	b = 0.1477)		
0	2.49258	2677.90	374.8224	1747.94	2.4811	8.563	0.00052
10	2.67295	3083.48	265.1182	1973.46	4.1869	7.437	0.00075
20	2.57032	3048.80	346.9072	1912.51	3.2755	7.521	0.00071
30	2.58087	3137.28	352.8593	1928.12	3.4267	7.309	0.00102
40	2.54609	3117.11	395.6781	1876.10	3.1364	7.357	0.00081
50	2.50664	3059.75	448.3412	1802.68	2.8038	7.494	0.00064
60	2.55187	3090.38	412.4067	1781.43	3.1739	7.420	0.00060
70	2.65417	3180.58	334.0567	1792.99	4.1538	7.210	0.00098
80	2.66125	3110.19	334.7157	1713.77	4.1720	7.373	0.00054
90	2.63561	2991.16	361.1341	1610.15	3.8241	7.666	0.00056
100	2.59967	2839.44	398.3286	1492.38	3.3818	8.076	0.00103
	PROD = 1 - 1400 bars			Average J = 2.588			
Water	Holser and Kennedy data			(a' = 116.174	b = 0.1477)		
120	2.59190	2596.34	418.3084	1298.59	3.1024	8.8321	0.00269
140	2.55180	2232.59	465.8947	1059.89	2.5171	10.2710	0.00106
160	2.53699	1924.74	493.8848	864.80	2.1461	11.9139	0.00070
180	2.45626	1566.51	630.5884	664.01	1.4312	14.6383	0.00066
200	2.50493	1341.76	569.2254	534.63	1.4179	17.0903	0.00079
220	2.49888	1084.66	602.8153	404.61	1.1281	21.1412	0.00078
240	2.50935	848.411	612.3082	294.91	0.9040	27.0281	0.00151
260	2.51420	620.861	636.1771	200.02	0.6615	36.9342	0.00179
280	2.55003	424.350	609.4099	125.92	0.4897	54.0380	0.00325
300	2.61224	249.222	545.2721	67.618	0.3331	92.0103	0.00452
320	2.73109	102.092	422.3950	25.103	0.1823	224.6109	0.00633
340	2.83728	-39.813	348.0263		-0.0892		0.00836
360	3.04543	-153.753	229.0061		-0.5404		0.01679
380§	3.44277	-247.801	106.4642		-1.9316		0.0749
400§	3.32964	-357.729	1205.544		-1.5444		7.139
	PROD = 1 - 1400 bars						

Average J (120 - 320°) = 2.551, (0 - 320°) = 2.569±0.069

(0 - 360°) = 2.60±0.128

§ Values past critical temperature

Table 8.2 (continued)

	Experimental quantities			Derived quantities			
T°K	J cc/mole	L atm	H atm $\times 10^{-4}$	P_i atm	$A \times 10^{+4}$	d Å	σ of v Eqn.8:25

Helium I (normal helium)   Keesom and Keesom data

T°K	J	L	H	P_i	A	d	σ
2.0	3.65895	10.02696	1.7806				0.0041
2.1	3.59147	9.41738	1.9732				0.0040
2.2	3.48656	8.63570	2.3534				0.0053
2.25	3.51105	8.75433	2.2664				0.0049
2.5	3.44910	7.96248	2.5458	*			0.0061
3.0	3.36354	6.12082	3.0272	*			0.0065
3.5	3.30656	3.98055	3.4194	*			0.0083
4.0	3.32937	1.97033	3.3015	*			0.0179
4.2	3.39220	1.32970	2.9298	*			0.0271

PROD vary, maximum pressure = 35 atm

Average $J \pm \sigma$ = 3.454±0.120

* See Chapter 9 for values

Helium II (superfluid helium)   Keesom and Keesom data

T°K	J	L	H	P_i	A	d	σ
1.25	3.72810	10.80988	1.82861				0.0026
1.50	4.14121	12.76589	1.01259				0.0087
1.75	4.43868	13.55201	0.68255				0.0092
1.80	3.17010†	17.4917†	0.3127†				–
1.90	5.5694†	15.8831†	0.3306†				–

PROD vary, maximum pressure = 35 atm

† The Deming method of statistical adjustment did not converge when used on the data for these two temperatures. Generally in such cases we discarded the whole set and did not report it. Here, however, we give instead the first approximation of J, L, and H as derived from the slope and intercept by least squares. The error in these terms is much greater than in the other sets. The whole set is so small that we felt we had to show these values; the derived quantities are not calculated.

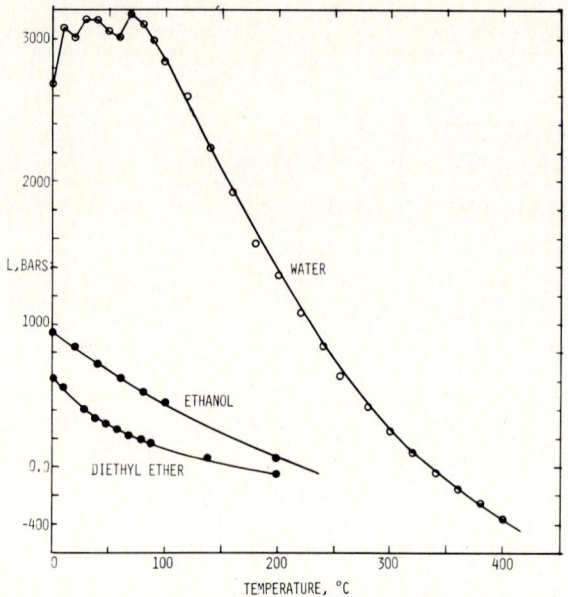

Fig.8.2. L, bars vs. temperature, °C.
Water – ○ points 100°C and below, Kennedy, Knight and Holser data, points 120°C and above, Holser and Kennedy data. ◐ points above critical point. Diethyl ether and ethanol – ● Amagat data.

AGREEMENT WITH EXPERIMENTAL DATA

We can now check to see how well the Tait-Tammann equation represents the data. The best procedure is to use the Tait-Tammann equation to calculate the volume and compare this calculated value with the experimental data. We must therefore integrate the Tait-Tammann equation. Writing this equation with separated variables, we have

$$\frac{dP}{L+P_e} = \frac{-dv}{J} \qquad (8:23)$$

Integrating, we have

$$\ln(P_e + L) = -v/J + \text{a constant} \qquad (8:24)$$

or, rewriting,

$$(P_e + L)e^{v/J} = H \qquad (8:25)$$

where H is the integration constant. H can be evaluated at each P_e and v, once J and L are known. H thus evaluated has different values for each pressure. We obtain a single value from the statistically different values of H by averaging the H over all the pressures. The values of J and L and the values of H thus obtained are the first approximation for use in the Deming iteration. Table 8.2 shows the values of J, L and H determined for a number of substances at various temperatures. Using these values and Eqn.8:25, we can determine the calculated volume. The values of v thus determined are in excellent agreement with the experimental values. Figure 8.3 gives the deviation of the calculated from the experimental values for diethyl ether. One must take into account the fact that the original data was given to only 4 digits, which makes the agreement remarkably good. The other substances give comparable results.

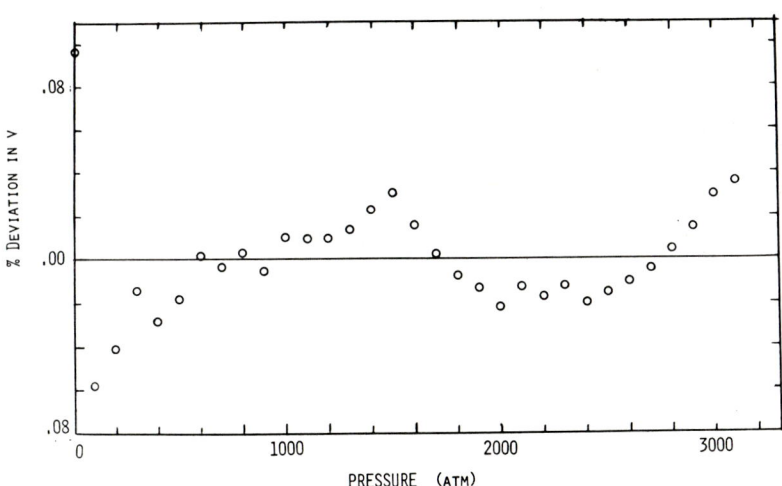

Fig.8.3. Diethyl ether. % deviation of calculated value of the volume from the actual volume vs. the pressure at 0°C.

VALUE OF P_i

We now have a method of calculating P_i, the internal pressure, from the Tait-Tammann coefficient, L, and the ratio, γ/a'. The first factor that we must investigate is the constancy of these parameters. Using Eqn.8:20 and Eqn.8:21, we have

$$Ae^{v/J} = \frac{L - P_i}{L + P_e} \qquad (8:26)$$

and inserting Eqn.8:25 we get

$$L - P_i = AH \qquad (8:27)$$

Since both A and H are integration constants and are constant at constant temperature and L is constant with temperature, then P_i must be constant with changes in pressure at constant temperature. Similarly, from Eqn.8:18, γ/a' is constant with pressure changes at constant temperature*. The surface tension, however, is not constant with a change in pressure, nor is a' constant under these changes.

In Table 8.2 are listed values of P_i for a number of substances. Generally, the values of P_i are less than the values currently in use. However, the current values are obtained using the ideal gas law and a very approximate calculation, whereas the values obtained here from the Tait-Tammann Law are free of approximation and use the association equation of state. An indication that these values are better can be obtained by calculating anew the thickness of the surface film, d, of the liquid, using Eqn.7:3 and the new values of P_i. These resulting values of d, which are given in Table 8.2 can be seen to be much larger than those calculated in Chapter 7. They are much more reasonable values and agree more closely with the estimate given by Rusanov. Examining the values of d, the surface film thickness, one sees that these values increase with increasing temperature. This is reasonable, since one would expect the region of the liquid that is less multiply-bonded to increase with an increase in the temperature (which increases the kinetic energy of the unimers). Since γ is not constant with pressure, d also changes with the application of pressure. This is, however, difficult to show experimentally.

---

* These constancies hold only as long as L is constant. There is some indication that at extremely high pressures, probably above 100.000 kg/cm^2, L is not constant with pressure. For the modifications necessary in the Tait-Tammann Law, see Kirkwood, Richardson and coworkers.

THE SURFACE AREA

The surface energy has been previously discussed in Chapter 7. There the concept of the surface-film volume (area), (B/d), and the thickness of the surface film were introduced. Now that we can calculate the exact values of B and d, it is interesting to compare the values of the surface energy as computed from $\gamma(B/d)$ with the former values based on the concept that $\gamma(Mv)^{2/3}$ is the surface energy. Table 8.3 presents such calculations. As can be seen from Table 8.3, the surface energy as calculated from B/d is many orders of magnitude greater than that calculated from $(Mv)^{2/3}$. The discrepancy is due to the surface area per formula weight as calculated from B/d being much greater. The question arises, which value is more correct or - better - more reasonable? We feel that it is entirely misleading to call $(Mv)^{2/3}$ the surface area, since the molar volume is not a cube of which one face is the surface. We feel that it is more logical to state that the surface film has a thickness and that the quantity we must deal with is the surface-film volume. Looking at such phenomena as the formation of colloids, films, or foams, we feel that it is more reasonable that the surface energy be equated to the association energy ($P_i B$) than that it be given as a quantity many orders of magnitude less, $\gamma(Mv)^{2/3}$. The stability of such colloids probably demands that the surface energy be much greater than the magnitude afforded by $\gamma(Mv)^{2/3}$.

VALUE OF THE COVOLUME B

The covolume, B, is another quantity of great interest. This quantity varies with temperature and pressure and hence is not given in Table 8.2. A plot of the values of B/v, a dimensionless quantity, is given for water in Fig.8.4, and for diethyl ether in Fig.8.5 (other substances behave similarly). Since liquids are relatively incompressible, the relative increase of the volume, v, with an increase in temperature is much greater than the decrease in volume with an increase in pressure. This behavior is relatively straightforward compared with the behavior of the covolume, B. Empirically B first decreases, passes through a minimum and then increases with an increase in the temperature at constant pressure. At constant temperature B

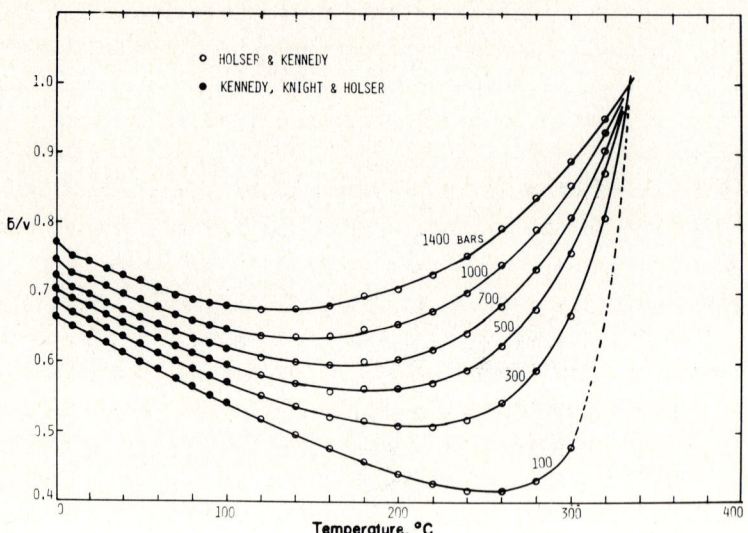

Fig.8.4. Water. E/v vs. $t°C$ for various pressures. The intercept at $E/v = 1.00$ is real (see Chapter 10).

Table 8.3

Comparison of surface energy computed by different methods

$t°C$	γ dynes/cm	$(Mv)^{2/3}$ cm^2	$\gamma(Mv)^{2/3}$ ergs	E cm^3	d ×10^{8}cm	E/d ×10^{-8}cm^2	$2\gamma(E/d)$ ×10^{-9}ergs
Water**	pressure = 0 bars						
0	75.83	6.88	521.91	11.79	8.56	1.38	20.87
10	74.35	6.88	511.82	11.56	7.44	1.55	23.11
20	72.88	6.89	502.15	11.35	7.52	1.51	21.99
30	71.40	6.90	492.77	11.14	7.31	1.52	21.77
40	69.92	6.92	483.71	10.95	7.36	1.49	20.82
50	68.45	6.94	474.84	10.77	7.49	1.44	19.66
60	66.97	6.96	466.14	10.59	7.42	1.43	19.11
70	65.49	6.99	457.52	10.41	7.21	1.44	18.91
80	64.01	7.01	449.04	10.24	7.37	1.39	17.78
90	62.54	7.05	441.33	10.07	7.67	1.31	16.43
100	61.06	7.08	432.29	9.90	8.08	1.23	14.97

Table 8.3 (continued)

t°c	γ $\frac{dynes}{cm}$	$(Mv)^{2/3}$ cm^2	$\gamma(Mv)^{2/3}$ ergs	E cm^3	d $\times 10^8 cm$	E/d $\times 10^{-8} cm^2$	$2\gamma(E/d)$ $\times 10^{-9}$ ergs
Methyl alcohol*	pressure = 0 atm						
0	24.05	11.61	279.18	21.06	8.48	2.48	11.94
9.2	23.34	11.69	272.76	20.64	8.78	2.35	10.97
23.85	22.21	11.81	262.34	19.96	9.58	2.08	9.25
30.05	21.73	11.94	259.51	19.85	10.16	1.95	8.49
Ethyl alcohol*	pressure = 0 atm						
0	24.05	14.84	356.79	29.38	9.80	3.00	14.42
20	22.39	15.05	336.87	27.92	10.96	2.55	11.42
40	20.72	15.27	316.36	26.43	12.84	2.06	8.53
n-Propyl alcohol*	pressure = 0 atm						
0	25.26	17.75	448.32	40.63	7.91	5.14	25.95
9.75	24.50	17.86	437.62	39.79	8.51	4.68	22.56
25.25	23.30	18.04	420.36	38.41	9.19	4.18	19.48
39.75	22.17	18.23	404.24	37.14	10.59	3.51	15.55
Water*	pressure = 0 atm						
0.0	75.83	6.88	521.63	11.78	8.60	1.37	20.77
5.0	75.09	6.88	516.48	11.66	7.25	1.61	24.15
10.0	74.35	6.88	511.51	11.55	8.00	1.44	21.47
15.0	73.61	6.88	506.66	11.44	7.86	1.46	21.43
20.0	72.88	6.89	501.96	11.34	7.46	1.52	22.16
30.0	71.40	6.90	492.55	11.14	6.99	1.59	22.78
40.0	69.92	6.91	483.43	10.94	6.99	1.57	21.89
50.0	68.45	6.93	474.60	10.76	7.17	1.50	20.54
80.0	64.01	7.01	448.71	10.23	9.21	1.11	14.22
100.0	61.06	7.08	432.02	9.89	8.95	1.11	13.49
198.0	46.59	7.59	353.61	8.38	16.11	0.52	4.84

* Amagat data
** Kennedy, Knight and Holser data

increases with an increase in pressure. This behavior is puzzling and part of it is explicable. Why should the covolume increase with an increase in pressure? The answer to this puzzle lies in Eqn.3:15 and the fact that $B \propto 1/\alpha$. As the pressure is increased, the normal structure of the α-mer is distorted and bonds are broken; hence α decreases and B increases.

On the other hand, with temperature increase, there is a more complex behavior. At lower pressures B first decreases, then increases. At higher pressures (see Fig.8.5) there appears to be only an increase in B; however, an examination of this figure shows that a minimum should occur at some lower temperature. A decrease in B is equivalent to an increase in α, and the converse is also true. Finally, all the curves intersect at $B/v = 1$. This intersection occurs, because L becomes equal to zero a number of degrees below the critical point. As a consequence, P_i also becomes equal to zero and, according to Eqn.8:19, B/v reduces to 1. This complex behavior is summarized for water and diethyl ether in Figs.8.4 and 8.5. A more complete discussion of the behavior of these parameters in the neighborhood of the critical point will be found in Chapter 10.

MOLECULAR VOLUME OF METHANOL

A very interesting facet of the B calculation is that it enables us to calculate some molecular parameters. The B for methanol at 0°C and 0 atm (extrapolated) is

$$B = 21.05887 \text{ cm}^3/\text{F.W.} = 34.98891 \text{ Å/unimer}$$

For a spherical unimer, the covolume $B = \frac{4}{3}\pi r^3$, where r is the radius of the covolume. Of course, a unimer in an α-mer is not a sphere but rather is bonded to its neighbors. Nevertheless, let us calculate the radius it would have if it were a sphere. Carrying out the calculation, $r = 2.029$ Å. Let us now calculate the covolume from molecular data. From Kennard's compilation of bond distances and bond angles derived from x-ray and electron diffraction experiments, we have

C—H distance	=	1.096 Å
C—O "	=	1.427 Å
O—H "	=	0.956 Å

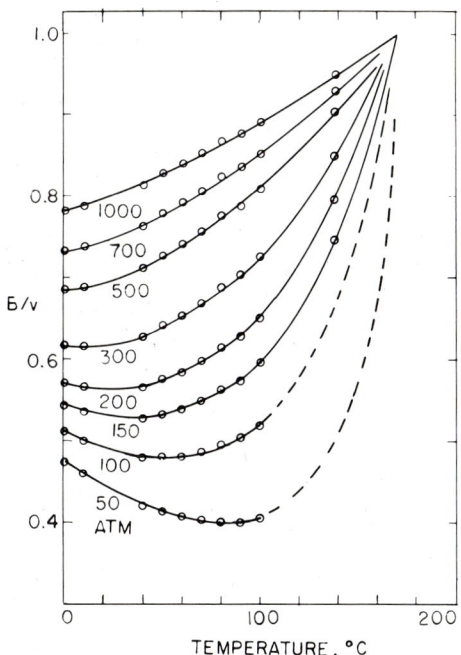

Fig.8.5. Diethyl ether. E/v vs. t°C for various pressures, Amagat data. The coincidence at $E/v = 1$ is a consequence of Eqn.8:19.

| H—C—H angles | = | 109.3° |
| C—O—H " | = | 108.9° |

To calculate the length of the unimer we shall assume the longest dimension to be that of the axis coincident with the C—O bond (see Fig.8.6). Since we need the projection of the C—H bond on this axis, we proceed thus. The methyl group forms a tetrahedron with a face angle of 1.096 Å. The values of the other quantities on the diagram by simple geometry and trigonometry are

$$k = a\cos\frac{HCH}{2} = 1.096 \cos 54.65° = 0.634112$$

$$j = 2a \sin\frac{HCH}{2} = 1.787868$$

$$s = 3j/2 = 2.681801$$

$$r = \sqrt{(s-j)^3/s} = 0.516113$$

$$h = \sqrt{k^2-r^2} = 0.368409$$

The projected distance of the C–H bond on the C–O axis is h. The projected length of the O–H bond on this same axis is

$$c = 0.956 \cos(180-COH) = 0.309665$$

The total lenght along this axis is then the sum of these three distances: h = 0.368409 Å
   C–O = 1.427
   c = 0.309665
   ──────────
Total length = 2.105074 Å

This value is about 3.7% larger than the assumed distance. Considering the gross assumptions of the unimer in the α-mer being a sphere of revolution, the agreement is not bad and primarily shows us that the B value is in the right range.

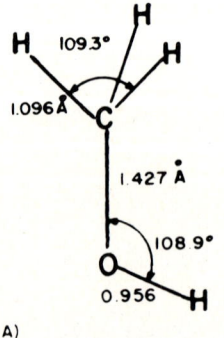
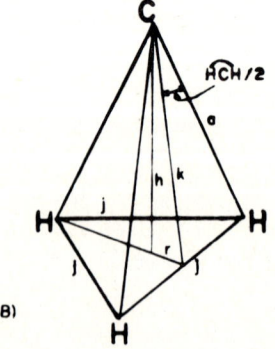

Fig.8.6. Methanol. (A) Bond distance and lengths from X-ray and electron diffraction studies. (B) Geometric relationships in the methyl group.

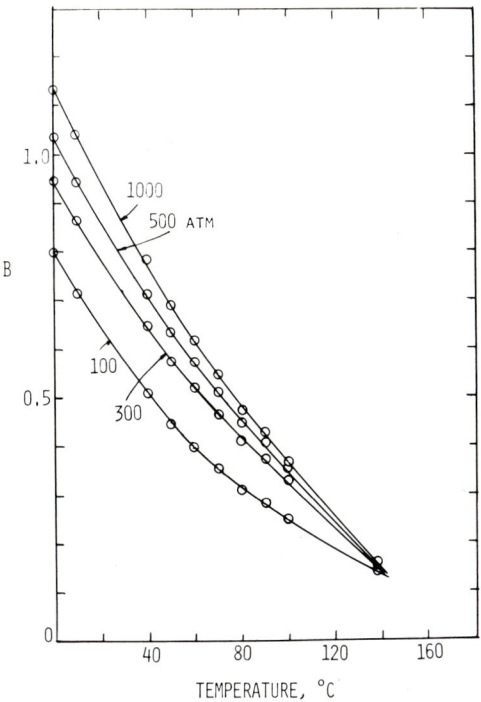

Fig.8.7. Variation of B with temperature and pressure for diethyl ether.

VALUE OF THE FACTOR B

The behavior of B with variation of the temperature and pressure is very interesting. In substances like diethyl ether, ethanol, and propanol, B increases with an increase in pressure and decreases with an increase in temperature, although the range of the variables over which these quantities are known is rather small (Fig.8.7). On the other hand, the behavior of B in water (Fig.8.8) with a variation of the same variables over about the same range, is quite different. Water (using the data of Kennedy et al.) exhibits a series of maxima and minima in the region below 100°C. These variations appear at all the plotted pressures and appear to be real and not errors in the data. Similar maxima and minima appear in the data of Amagat when they are plotted, although they are not as extensive or complete as the data of Kennedy et al. Water in general behaves anomalously,

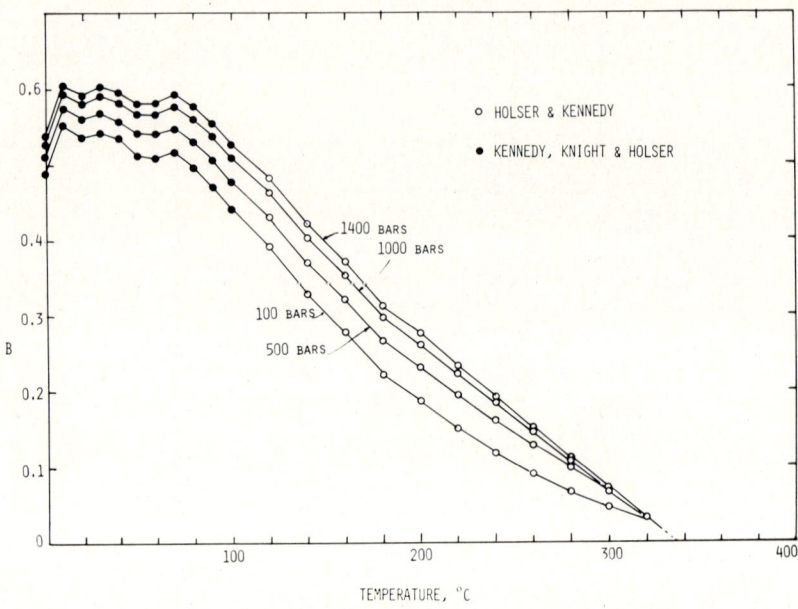

Fig.8.8. Water. B vs. t°C for various pressures. See Chapter 10 for discussion of point at zero.

especially around 4°C. But apparently its peculiar behavior extends beyond the maximum in density at 4°C. Irregularities have been noted at higher temperatures, but it is difficult to pin down exactly either the temperatures at which these variations occur or the exact reason why they happen (see for example the work of Drost-Hansen). Such variations in the behavior of water exist not only in the behavior of B but also in the variation of L and hence of P_i (see Fig.8.2). It seems entirely reasonable that irregularities in structure should be reflected in abnormalities in the value of B, the number of bonds broken, and in P_i, the internal pressure or the pressure exhibited by the close-packed species. This variable offers us a valuable tool in such investigations.

IMPORTANT EQUATIONS

Tait-Tammann Equation

$$-\left(\frac{\partial v}{\partial P_e}\right)_T = \frac{J}{L + P_e} \qquad (8:6)$$

Integrated Tait-Tammann Equation

$$(P_e + L)e^{v/J} = H \qquad (8:25)$$

$$v = J \ln[H/(L + P_e)] \qquad (8:25)$$

$$(1 - E/v) = Ae^{v/J} \qquad (8:21)$$

$$P_i = \frac{\gamma}{a'} L \qquad (8:18)$$

$$\frac{E}{v} = \frac{(P_e + P_i)}{(P_e + L)} = \frac{(P_e + P_i)}{(a' P_i/\gamma) + P_e} \qquad (8:19)$$

$$L - HA = P_i \qquad (8:27)$$

REFERENCES

Amagat, E.H., *Mémoires sur l'élasticité et la dilation des fluides jusqu'aux très hautes pressions*, Ann. Chim. Phys., 29 (1893) 503-574.

Brown, S.D. and Ginell, R., *The partial structure of glass and the reconstructive nature of devitrification processes*, Symposium on Nucleation and Crystallization on Glasses and Melts, (Eds. M.K. Reser, G. Smith, and H. Insley) Am. Ceram. Soc., 1962, pp.109-118.

Carl, H., *Prüfung der Kompressionsgleichung der Flüssigkeiten an den Daten von Amagat und Bridgman*, Z. Phys. Chem. (Leipzig), 101 (1922) 238-268.

Deming, W.E., *Statistical Adjustment of Data*, Wiley, New York, 1945, reprinted by Dover Publications, 1964.

Drost-Hansen, W., *Anomalies in the properties of water*, Proc. 1st Int. Symp. Water Desalination, Washington, D.C., 1965, Vol.1, pp.382-406 (Publ.1967)

Gibson, R.E. and Loeffler, O.H., *Pressure-volume-temperature relations in solution. II: The energy-volume coefficients of aniline, nitrobenzene, bromobenzene and chlorobenzene*, J. Amer. Chem. Soc., 61 (1939) 2515-2522.

III. Some thermodynamic properties of mixtures of aniline and nitrobenzene, ibid., 61 (1939) 2877-2884.

V: *The energy-volume coefficient of carbon tetrachloride, water, and ethylene glycol*, ibid., 63 (1941) 2287-2295.

VIII: *Water solutions of sodium chloride and sodium bromide*, Ann. N.Y. Acad. Sci., 51 (1941) 727.

Ginell, R., *Tait coefficients and λ transition of helium I and II*, J. Chem. Phys., 35 (1961) 471-478.

Ginell, R. and Quigley, T.J., *Compressibility of solids and Tait's Law: I: p-v relationships of the alkali metals*, J. Phys. Chem. Solids, 25 (1965) 1157-1169.

Hayward, A.T.J., *Compressibility equation for liquids - a comparative study*, Natl. Eng. Lab. Report 295, East Kilbride, Glasgow, 1967.

Hirschfelder, J.O., Curtiss, C.F. and Bird, R.B., *Molecular Theory of Gases and Liquids*, Wiley, New York, 1954.

Holser, W.T. and Kennedy, G.C., *Properties of water. Part IV: Pressure-volume-temperature relations of water in the range 100-400°C and 100-1400 bars*, Am. J. Sci., 256 (1958) 744-753.

Keesom, W.H. and Keesom, A.P., *Thermodynamic diagrams of liquid helium*, Physica, 1 (1934) 128-133.

Kennard, O., *Bond lengths between carbon and other elements*, Handbook of Chem. and Physics, 51st edn., 1970-71, R.C.Weast, Ed., Chem. Rubber Co., Cleveland, Ohio.

Kennedy, G.C., Knight, W.L. and Holser, W.T., *Properties of water. Part III: Pressure-volume-temperature relations of water in the range 0-100°C and 1-1400 bars*, Am. J. Sci., 256 (1958) 590-595.

For high pressure modifications of the Tait equation, see

Kirkwood, J.G. and Bethe, H., *The pressure wave produced by an underwater explosion, Part I*, OSRD Rept. 588 (Dept. of Comm. No. PB 32182).

Kirkwood, J.G. and Richardson, J.M., *The pressure wave produced by an underwater explosion, Part III*, OSRD Rept. 813 (Dept. of Comm. No. PB 32184).

Richardson, J.M., Arons, A.B. and Halverson, R.R., *Hydrodynamic properties of sea water at the front of a shock wave*, J. Chem. Phys., 15 (1947) 785-794.

MacDonald, J.R., *Some simple isothermal equations of state*, Rev. Mod. Phys., 38 (1966) 669-679.

Rusanov, A.I., *Recent investigations on the thickness of surface layers*, Prog. Surf. Membr. Sci., 4 (1971) 57-114.

Tait, P.G., *Collected Scientific Papers*, Cambridge University Press, Cambridge, Vol.II, 1893.

Tammann, G., *Über die Beziehung zwischen den inneren Kräften und Eigenschaften der Lösungen*, Leopold Voss, Hamburg, 1907, p.38.

Wohl, A., *Untersuchungen über die Zustandsgleichung*, Z. Phys. Chem. (Leipzig), 99 (1921) 207-241.

CHAPTER 9

THE LIQUID AND THE SOLID STATE

The picture of the liquid state as presented in Chapter 8 is simple and seems satisfactory. Although there is nothing wrong with the Tait-Tammann equation, there are still complications.

THE SATURATED VOLUME

As an example, let us use the values at the saturation point. Here the saturation pressure and the saturated volume are well known quantities for many substances. We rewrite Eqn. 8:22 to give

$$v = J \cdot \ln[H/(L + P_e)] \qquad (9:1)$$

Now, setting the saturation pressure equal to P_e, we can calculate the saturated volume. The results of this calculation for several substances are given in Table 9.1. We give both the volume per formula weight and the density. The latter is the usual quantity tabulated. As can be seen from the table, the deviation of the theoretical density from the experimental density is very small. The calculation depends to a large degree upon the value of the Tait-Tammann coefficients and not as much on the value of the saturation pressure. The utility of this determination is clear. If we know the saturation pressure and the saturated volume, we have a method of checking whether the Tait-Tammann coefficients are valid. This means that we have a method of checking whether high pressure values such as compressions are good data.

It would be interesting to see, now that we can calculate the saturated volume from the saturation pressure, whether we can reverse the process and calculate the saturation pressure from the saturated volume. Algebraically, nothing could be simpler than inverting the equation, but practical realities make it almost impossible to reverse the calculation. The reversed equation involves the difference between two large quantities, one of which depends on an exponential term involving v. Slight variations in v change the value of the exponential term and hence the value of the difference, making the calculation extremely difficult to carry out accurately. An example of such a calculation is given in Table 9.2.

DEGREE OF AGGREGATION AT THE SATURATION POINT

We are now in a position to calculate the degree of aggregation at the saturation point. We solve Eqn.7:1 for B, to get

$$B = \frac{(P_e + P_i)(v-B)}{N_w RT} \tag{9:2}$$

Using Eqn.8:7 in the form

$$(v-B) = Ave^{v/J} \tag{9:3}$$

we have

$$B = \frac{(P_e + P_i)Ave^{v/J}}{N_w RT} \tag{9:4}$$

As defined in Chapter 2, the degree of aggregation is

$$Z = \frac{\sum xN_x}{\sum N_x} = \frac{w/M_0}{B} \tag{9:5}$$

Since in these calculations we set w = FW, the formula weight, and since we take M_0 = FW, Eqn.9:5 reduces to $Z = 1/B$. We can readily calculate B and Z, since all the factors have previously been determined. In Table 9.3 we report these results for a few substances. We also list in this table the values of the degree of aggregation of the saturated vapor. This value can be obtained using the ideal gas law*.

---

* The use of this law is justified here, since the gas is so attenuated that the free volume, $(v-B)$, is essentially equal to the actual volume, v. For instance, at 0°C the molar volume of water is about 200 times as great as the Normal Molar Volume, 22.414 liters.

Table 9.1

Comparison of the theoretical density, ρ, and the volume per formula weight, v, at the saturation point; calculated from the Tait-Tammann coefficients and the saturation pressure, using the experimental values for various substances.

t°C	P_{sat} atm*	volume cc/FW Theor.	Exp.	density g/cc Theor.	Exp.*	% dev. in ρ [§]
Diethyl ether (liq.)						
0	0.2433	100.6687	100.683	0.73631	0.7362	+0.01
10	0.3826	102.2244	102.267	0.72510	0.7248	+0.04
40	1.2118	107.4473	107.518	0.68985	0.6894	+0.07
50	1.6789	109.4913	109.585	0.67698	0.6764	+0.09
60	2.2816	111.5991	111.329	0.66419	0.6658	−0.24
70	3.0316	113.9058	113.477	0.65074	0.6532	−0.38
80	3.9132	116.3762	115.781	0.63693	0.6402	−0.51
90	5.0408	118.9521	118.597	0.62313	0.6250	−0.30
100	6.3882	121.8060	121.414	0.60853	0.6105	−0.32
138	14.0097	135.9873	136.639	0.54507	0.5425	+0.47
Methyl alcohol (liq.)						
0	0.03895	39.5472	39.5585	0.81023	0.8100	−0.03
9.2	0.06859	39.9469	39.977	0.80210	0.8015[†]	−0.08
23.85	0.15436	40.6010	40.662	0.78918	0.7880[†]	−0.15
30.05	0.21106	41.2648	40.948	0.77649	0.7825[†]	−0.76
Ethyl alcohol (liq.)						
0	0.01611	57.1414	57.1400	0.80623	0.80625	−0.00
20	0.05790	58.3590	58.3597	0.78941	0.7894	+0.00
40	0.17670	59.6621	59.6596	0.77217	0.7722	−0.00
60	0.46070	61.0975	61.0916	0.75403	0.7541	−0.01
80	1.05495	62.6982	62.6962	0.73478	0.7348	−0.00
100	3.10000[†]	64.5000	64.3693	0.71425	0.7157[†]	−0.20
198	28.04424	81.8779	82.0481	0.56266	0.5615	+0.21

* Young's values from Timmermans
\# Values from steam tables
§ % dev. = $\dfrac{\rho(\text{theor.}) - \rho(\text{exp.})}{\rho(\text{exp.})}$
** Data from Kennedy et al.
† interpolated values

Table 9.1 (continued)

t°C	P_{sat} bars*	volume cc/FW Theor.	Exp.	density g/cc Theor.	Exp.**	% dev. in ρ §
Water						
0	0.00611	18.0563	18.0580	0.99988	0.99977	+0.01
10	0.01228	18.0603	18.0591	0.99965	0.99972	−0.01
20	0.02338	18.0871	18.0850	0.99817	0.99828	−0.01
30	0.04243	18.1313	18.1312	0.99574	0.99574	−0.00
40	0.07376	18.1950	18.1955	0.99225	0.99222	+0.00
50	0.12334	18.2728	18.2733	0.98802	0.98800	+0.00
60	0.19916	18.3639	18.3634	0.98313	0.98315	−0.00
70	0.31157	18.4644	18.4660	0.97777	0.97769	+0.01
80	0.47343	18.5782	18.5809	0.97178	0.97164	+0.01
90	0.70095	18.7022	18.7072	0.96534	0.96508	+0.03
100	1.01325	18.8370	18.8436	0.95843	0.95810	+0.03
120	1.98535	19.1384	19.1490	0.94334	0.94282	+0.06
140	3.61425	19.5002	19.4984	0.92584	0.92592	−0.01
160	6.18081	19.9075	19.9041	0.90689	0.90705	−0.02
180	10.02608	20.3722	20.3550	0.88621	0.88696	−0.09
200	15.54423	20.8946	20.8847	0.86405	0.86446	−0.05
220	23.19222	21.4948	21.4820	0.83992	0.84042	−0.06
240	33.46452	22.1966	22.1921	0.81337	0.81353	−0.02
260	46.91335	23.0348	23.0261	0.78377	0.78407	−0.04
280	64.13348	24.0506	24.0518	0.75067	0.75063	+0.01
300	85.90310	25.3312	25.3479	0.71272	0.71225	+0.07
320	112.90614	26.9986	27.0723	0.66870	0.66688	+0.27

T°K	P_{sat} mm Hg #∇	volume cc/FW Theor.	Exp.##	density g/cc Theor.	Exp.∇∇	% dev. in ρ §
Normal helium 4						
2.5	77.490	27.79	27.62	0.1440	0.1449	−0.62
3.0	182.073	28.48	28.32	0.1405	0.1413	−0.57
3.5	355.849	29.58	29.49	0.1353	0.1357	−0.29
4.0	616.537	31.24	31.14	0.1281	0.1285	−0.31
4.2	749.128	32.04	31.97	0.1249	0.1252	−0.24

\# Data from Keesom
\#\# Data from Kerr and Taylor
∇ Temp. Scale = T_{58}
∇∇ calculated from volume

Table 9.2

Values of the pressure of saturated diethyl ether, calculated by using the Tait-Tammann equation* and various densities close to the reported saturated density. Temperature = 60°C

v cc/FW	ρ	P_e atm
111.00	0.66777	20.713
111.30	0.66597	11.347
111.50	0.66478	5.256
111.60	0.66418	2.255
111.70	0.66359	-0.717

From the literature ** P_{sat} (60°C) = 2.28158 atm = 1734 mm Hg
v_{sat} (60°C) = 111.329 cc/FW
ρ_{sat} (60°C) = 0.6658

* $P_e = He^{-v/J} - L$
** Data from Timmermans

If one examines Table 9.3, certain features seem very strange. These can be listed. The values for Z of the liquid seem in general very low. The values for diethyl ether, which appears to be a simple liquid, are of the same order of magnitude as those of water, which has a high dipole moment and which we normally expect to be highly associated. For temperatures below 100°C diethyl ether gives greater values for the degree of association, Z, than does water. In the vapor state the values of Z for water fall below 1 up to above 30°C.

To try to rationalize these strange results let us examine the fact that the value of Z in water vapor below about 40°C are below 1. We would normally expect a vapor to consist of predominantly 1-mers with a small amount of 2-mers and a much smaller number of any of the higher j-mers. A mixture of this sort would be expected to yield a Z value slightly greater than 1. That the Z value lies below 1 points to the fact that this is not a "normal" case. Perhaps the difficulty lies in the fact that we used the ideal gas law to determine $B_{vap} = (\sum N_x)_{vap}$? We have neglected several factors in using this law. Let us examine them serially. First, the covolume, B, has been ignored, setting

$(v-B) = v$, i.e., the free volume equal to the actual volume. However, this cannot be the trouble, since $(v-B)$ is smaller than v and its use would have the effect of making the value of B_{vap} smaller. Second, there is no internal pressure term in P, i.e., $P = P_e \neq (P_e + P_i)$. However, generally, we have not used P_i in describing the vapor state and we do not think it is necessary here. The only conclusion is that our primitive concept of the vapor as consisting of 1-mers, 2-mers and higher j-mers is too elementary. Even if all the vapor consisted of 1-mers, which implies a total absence of binary collisions, the Z would be equal to 1 and not be below 1! A definitely new concept must be introduced. Since we know that dissociation occurs in the liquid state and that at saturation the liquid is in equilibrium with the vapor, we make the assumption that dissociation also occurs in the vapor state. Put mathematically,

$$B' = \sum N_x' = gN_1' + N_1 + N_2 + N_3 + \ldots \qquad (9:6)$$

where N_1' is the number of moles of N_1 that are dissociated and g is the number of particles into which each N_1' dissociates. B' is the total number of particles.

For water the amount of dissociation in the liquid state is an accessible quantity. We know that in water liquid there is ionization into the hydrogen ion, H^+, and the hydroxide ion, OH^-. In the liquid the hydrogen ion is probably complexed to a neutral water molecule, giving a hydronium ion, H_3O^+. Since the liquid water is in equilibrium with the water vapor, we shall make the assumption that the concentrations of the 1-mers, the H^+, and the OH^- in the liquid are equal to that in the vapor. Since tables of pK_w, the ionization constant of liquid water, are available in the literature we can calculate the value of C_1', the concentration of the dissociated water in the liquid. Thus

$$pK_w = -\log_{10}(?C_1')$$

and

$$(N_1')_{vap} = (C_1') \cdot v_{vap}$$

We take the concentration in the liquid to be equal to the concentration in the vapor. We now calculate

$$B' - gN_1' = N_1 + N_2 + N_3 + \ldots \qquad (9:7)$$

Table 9.3

Values of B and Z at saturation point

t°C	P_i bars*	P_{sat} bars**	B_{liq}	Z_{liq}	Z_{vap}
Water					
0	1747.94	0.006105	0.4830	2.0704	0.99878
10	1973.46	0.01228	0.5447	1.8358	0.99794
20	1912.51	0.02338	0.5285	1.8922	0.99842
30	1928.12	0.04243	0.5346	1.8707	0.99920
40	1876.10	0.07376	0.5221	1.9153	1.00030
50	1802.68	0.12334	0.5038	1.9849	1.00171
60	1781.43	0.19916	0.5002	1.9994	1.00326
70	1792.99	0.31157	0.5066	1.9738	1.00514
80	1713.77	0.47343	0.4875	2.0514	1.00765
90	1610.15	0.70095	0.4616	2.1665	1.01043
100	1492.38	1.01324	0.4312	2.3189	1.01374
120	1298.59	1.9854	0.3813	2.6224	1.02494
140	1059.89	3.6143	0.3163	3.1611	1.03528
160	864.80	6.1808	0.2639	3.7895	1.05208
180	664.01	10.0261	0.2070	4.8308	1,07401
200	534.63	15.5442	0.1730	5.7802	1.10239
220	404.61	23.1922	0.1369	7.3056	1.13815
240	294.91	33.4645	0.1070	9.3447	1.18378
260	200.02	46.9133	0.08056	12.413	1.24215
280	125.92	64.133	0.06075	16.461	1.31848
300	67.618	85.903	0.04454	22.454	1.41932
320	25.103	112.91	0.02787	35.879	1.56293

t°C	P_i atm*	P_{sat} atm***	B_{liq}	Z_{liq}	Z_{vap}
Methyl alcohol					
0	559.80	0.03895	0.4632	2.1588	1.0094
9.2	524.90	0.06859	0.4409	2.2678	1.0035
23.85	457.19	0.15436	0.3938	2.5391	1.0131
30.05	422.80	0.21106	0.3350	2.9849	1.0224

Table 9.3 (continued)

t°C	P_i atm*	P_{sat} atm***	B_{liq}	Z_{liq}	Z_{vap}
Ethyl alcohol					
0	484.30	0.01611	0.5997	1.6675	0.9969
20	403.03	0.05789	0.5100	1.9610	1.0011
40	318.55	0.17670	0.4119	2.4279	0.9943
60	252.20	0.46070	0.3340	2.9942	1.0175
80	194.06	1.0550	0.2645	3.7802	1.0375
100	153.23	2.8044	0.2120	4.7168	0.8319
198	9.661	3.1000	0.04665	21.4341	1.3083
Diethyl ether					
0	269.78	0.2433	0.6889	1.4516	1.0279
10	231.28	0.3826	0.6019	1.6613	1.0355
40	139.58	1.2118	0.3847	2.5993	1.0673
50	112.76	1.6789	0.3188	3.1364	1.0822
60	92.53	2.2816	0.2582	3.8734	1.0945
80	58.59	3.9132	0.1700	5.8815	1.1539
90	47.72	5.0408	0.1494	6.6931	1.1779
100	36.71	6.3882	0.1230	8.1275	1.2073
138	1.197	14.0097	0.0521	19.1951	1.3981

* see cautionary note in Chapter 8 (page 130)
** from steam tables in handbook
*** from Timmermans

for water, since we know that $g=2$. We ignore here the possibility that the H_3O^+ forms, since it would enter only as a small correction. We know that for a formula weight of vapor

$$\sum x N_x = 1$$

so that

$$\sum x N_x - N_1' = 1 - N_1' = N_1 + 2N_2 + 3N_3 + 4N_4 + \ldots \tag{9:8}$$

The term g does not enter in Eqn.9:8, since $\sum x N_x$ is concerned with the number of unimers rather than with the number of particles. Subtracting Eqn.9:7 from Eqn.9:8 gives for water

$$(1-N_1')-(B'-2N_1') = N_2+2N_3+3N_4+\ldots = M_x \qquad (9:9)$$

As a rough approximation we can consider that

$$M_x \simeq N_2$$

since the concentration of the higher j-mers in the vapor is very small at these low temperatures and highly attenuated volumes. Hence we can calculate the approximate number of N_1 in the vapor by

$$(B'-2N_1') - M_x = M_z \simeq N_1 \qquad (9:10)$$

We can calculate the degree of dissociation of water excluding the dissociated particles by computing

$$Z''_{vap} = 1/(B'-2N_1')$$

where Z'' is the degree of association after subtracting the effect of dissociation. The result of these calculations for water from 0°C to 60°C is shown in Table 9.4. As can be seen from the table, these calculations explain the below-1 values in the vapor phase. If we now examine the values of ethyl alcohol in Table 9.3, we realize that this substance must also be somewhat dissociated in the vapor and liquid states. This dissociation has been postulated for ethyl alcohol in the liquid state by acid-base theories.

Some of the difficulties with the values have thus been cleared up, but not all of them. If we examine the values of Z'' for water in Table 9.4, we see that the values, while positive, are decreasing with an increase in temperature, while the values for diethyl ether are increasing. If one compares these values of Z'' with the values of Z in Table 9.3, we see that there is an anomaly in the behavior of water. The values in Table 9.3 oscillate and these oscillations are real (see Chapter 8). It is coincidental that our pK_w values only go to 60°C. Hence it would be expected that these values of Z would behave like those of diethyl ether and increase after 60°C. The cause of this decrease undoubtedly lies in the detailed structure of water, which still remains to be solved. It would be very interesting if we had values from which the dissociation of other substances could be calculated. The dissociation of many substances has been postulated on theoretical grounds but hard evidence to substantiate such claims is difficult to obtain.

Table 9.4
Calculation of the degree of aggregation of water in the vapor state assuming dissociation occurs

t°C	$-\log_{10}X_w$*	$[H^+] \cdot 10^8$ $= C_1'$	v_{sat} cc/FW	N_1'	B	$M_x^\dagger$	$M_z^{\dagger\dagger}$	$C_z \cdot 10^{7\nabla}$	Z_{vap}	$Z''^{\nabla\nabla}$
0	14.9435	3.3748	3724642.8	0.1257	1.00122	0.1245	0.6253	1.6789	0.999	1.334
10	14.5346	5.4038	1921436.5	0.1038	1.00201	0.1018	0.6925	3.6042	0.998	1.259
20	14.1669	8.2518	1044234.4	0.0862	1.00159	0.0846	0.7447	7.1312	0.998	1.206
30	13.8330	12.120	594531.75	0.0721	1.00080	0.0713	0.7854	13.210	0.999	1.167
40	13.5348	17.084	352887.00	0.0603	0.99970	0.0606	0.8185	23.195	1.000	1.137
50	13.2617	23.396	217469.01	0.0509	0.99829	0.0526	0.8439	38.807	1.002	1.115
60	13.0171	31.006	138630.15	0.0430	0.99675	0.0462	0.8646	62.363	1.003	1.098

* from handbook

$\dagger$ $M_x = N_2 + 2N_3 + 3N_4 + 4N_5 + \cdots \approx N_2$

$\dagger\dagger$ $M_z = (B' - 2N_1) - M_x \approx N_1$

∇ $C_z = M_z/v_{vap} \approx C_1$

$\nabla\nabla$ $Z'' = 1/(B - M_z)$

We still have much to learn about the structure of liquids, and data on dissociation would go far to enable us to eliminate many points of ignorance.

THE MAXIMUM IN B

In many ways B is a complicated quantity. Another aspect of its complexity is its behavior as the pressure increases. If we plot the value of B as a function of the pressure at constant temperature, we find that the curve exhibits a maximum (see Fig.9.1). However, to find the value of B at the maximum, it is more convenient to proceed analytically than graphically. Taking the partial derivative of Eqn.9:4 with respect to P_e (T = constant), we have

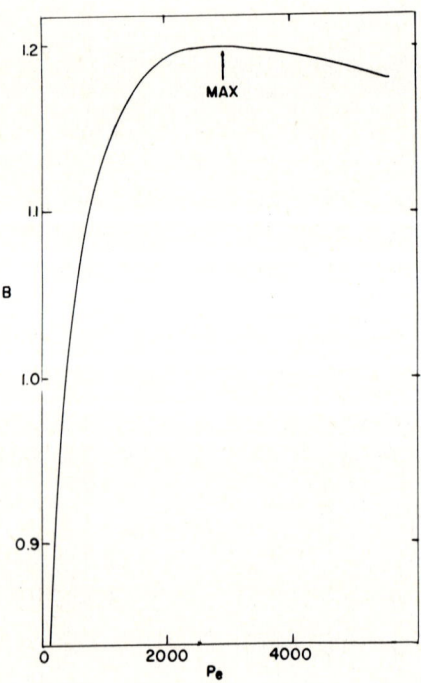

Fig.9.1. Diethyl ether. B vs. P_e.

$$\left(\frac{\partial B}{\partial P_e}\right)_T = \frac{A e^{v/J}}{N_w RT}\left[v + (P_e + P_i)\left(1 + \frac{v}{J}\right)\frac{\partial v}{\partial P_e}\right] \quad (9:11)$$

Inserting the Tait-Tammann equation (Eqn.8:6), we obtain

$$\left(\frac{\partial B}{\partial P_e}\right)_T = \frac{A e^{v/J}}{N_w RT}\left[v + (P_e + P_i)\left(1 + \frac{v}{J}\right)\left(\frac{-J}{L+P_e}\right)\right] \quad (9:12)$$

Since at the maximum $\partial B/\partial P_e = 0$, then, equating the right-hand side of Eqn.9:12 to zero and simplifying, we have

$$0 = \frac{v}{J}(L + P_e) - (P_e + P_i)(1 + v/J) \quad (9:13)$$

or

$$P_e = (L - P_i)(v/J) - P_i \quad (9:14)$$

The same equation is obtained by differentiating with respect to v. Using this equation together with Eqn.8:22

$$\frac{v}{J} = \ln[H/(L + P_e)] \quad (8:22)$$

we can solve for P_e iteratively using known values of P_i and the Tait-Tammann coefficients. We then can obtain B_{max} using Eqn.9:4. The values from this calculation for several substances will be found in Table 9.5. When using the water data of Kennedy et al. we cannot calculate values for temperatures at which L is negative, since a negative L causes the iteration to break down and gives rise to a region of singularity (for more details on this region see Chapter 10). Also given in this table is Z at the maximum in B. Since $Z = 1/B$ the maximum in B is transformed into a minimum in Z.

We must now ask, why is there a maximum in B, the number of bonds broken, or conversely, a minimum in Z, the degree of aggregation? When a liquid is compressed, two factors cause changes in its structure: the first is that the compression causes the unimers in the α-mer to move in order to accomodate the increase in pressure. This movement, which can be considered an illustration of the working of Le Chatelier's Law, results in breaking the structure or equivalently in breaking bonds. The second factor is that, because the structure is broken and distorted to accomodate the pressure increase, the voids that are normally present in the liquid are partially filled, thus causing the

Table 9.5
Values of B_{max} and Z_{min}§

t°C	P_i bars	P_e bars	v cc/FW	B_{max}	Z_{min}
Water*					
0	1747.94	4122.10	15.733	0.55615	1.79807
10	1973.46	4524.12	15.646	0.63009	1.58708
20	1912.51	4982.79	15.597	0.62427	1.60186
30	1928.12	5361.53	15.559	0.64023	1.56194
40	1876.10	5701.85	15.547	0.63677	1.57044
50	1802.68	5997.00	15.553	0.62667	1.59573
60	1781.43	6197.74	15.556	0.63151	1.58351
70	1792.99	6339.03	15.555	0.64624	1.54741
80	1713.77	6464.69	15.586	0.63315	1.57940
90	1610.15	6583.04	15.636	0.61203	1.63391
100	1492.38	6644.24	15.703	0.58495	1.70955
120	1298.59	6638.24	15.852	0.53982	1.85247
140	1059.89	6327.48	16.075	0.47360	2.11148
160	864.799	5961.02	16.338	0.41622	2.40259
180	664.011	5471.12	16.698	0.34868	2.86797
200	534.627	4958.87	17.049	0.30499	3.27881
220	404.605	4362.01	17.515	0.25423	3.93343
240	294.906	3693.86	18.083	0.20601	4.85406
260	200.023	2950.05	18.819	0.15761	6.34477
280	125.923	2188.29	19.775	0.11366	8.79830
300	67.618	1403.62	21.163	0.07179	13.92985
320	25.103	640.55	23.613	0.03304	30.26510
340#					
360#					

t°C	P_i atm	P_e atm	v cc/FW	B_{max}	Z_{min}
Methyl alcohol†					
0	559.80	3581.94	33.628	0.65916	1.51707
9.2	524.90	3534.69	33.816	0.63917	1.56453
23.85	457.52	3432.43	34.180	0.59147	1.69070
30.05	422.80	3259.37	34.631	0.56369	1.77402

Table 9.5 (continued)

t°C	P_i atm	P_e atm	v cc/FW	B_{max}	Z_{min}
Ethyl alcohol[††]					
0	484.30	3789.67	48.721	0.89858	1.11287
20	403.03	3618.68	49.372	0.81256	1.23068
40	318.55	3389.17	50.211	0.70632	1.41578
60	252.20	3124.16	51.110	0.61857	1.61663
80	194.06	2815.36	52.166	0.53215	1.87917
100	153.23	2515.85	53.231	0.47226	2.11749
198	96.605	446.149	68.680	0.08002	12.49698
Diethyl ether[†]					
0	269.78	2948.23	84.469	1.2008	0.83276
10	231.28	2770.24	85.487	1.0943	0.91381
40	139.58	2202.26	88.987	0.80933	1.23559
50	112.76	1981.83	90.510	0.70765	1.41312
60	92.533	1794.04	91.998	0.62986	1.58766
70	74.435	1599.62	93.671	0.55310	1.80799
80	58.585	1407.66	95.535	0.47873	2.08887
90	47.717	1257.10	97.276	0.42969	2.32728
100	36.708	1088.77	99.469	0.36915	2.70894
138	1.1968	551.832	110.386	0.18164	5.50527

* Kennedy et al. data
[†] Amagat data, first series
[††] Amagat data, second series
\# Iterative solution not possible since L is negative
§ Assuming boundary condition $P_e = 0$ can be used (see Chapter 8)

formation of new bonds. The number of bonds broken at the beginning of the process is greater than the number of bonds formed. The reason behind this statement is the pragmatic fact that the form at lower pressures is apparently more stable than the form at higher pressures, i.e., the structure at lower pressures must have more bonds than the form at higher pressures. When the pressure is released, the structure reverts spontaneously to its low pressure structure. When the pressure reaches a high enough level, the number of bonds formed exceeds the number of bonds broken. The result of these two factors is a maximum in B and conversely a minimum in Z. With some substances the resulting high pressure structure becomes a metastable form which has an appreciable lifetime.

In the case of methyl alcohol, examining the table, we see that Z approaches 1 at the minimum, the degree of aggregation of the 1-mer. On the other hand, in the case of diethyl ether, the value of Z (at 0°C) at the minimum shows that there must be dissociation at very high pressures. This presents the possibility that if we subjected mixed ethers to high pressures there would be metathetical changes, which might be of some interest to synthetically oriented chemists.

THE VALUES OF NORMAL HELIUM 4

Helium 4 is known to have peculiar properties and thus it is not surprising to find that its behavior here is also different. The values of the surface tension of helium 4 are given by Ure, Keesom and Onnes and by Allen and Misener. Between 2.5°K and 4.2°K the surface tension appears to give a straight line. Hence we have fitted a least square straight line to this portion. From the coefficients thus obtained we can calculate P_i by the use of Eqn.8:18. Using these values of P_i we can then calculate values of P_e at the maximum in B using Eqn.9:14 and Eqn.8:22 as done in the previous section. If we do this, we find that, although the process converges, we obtain negative values for P_e. This is a little curious, since it is very difficult to ascribe meaning to negative P_e. Apparently, whatever $P_e(max)$ is, it is not very large. The lowest pressure possible in a liquid system is the saturation pressure, so for want of a better assumption we assume that the maximum in B occurs at a pressure equal to that at saturation. The meaning of this assumption is that, although helium liquid is in 5-symmetry because it is a liquid, nevertheless, when it is compressed the number of new bonds formed exceeds the number of bonds broken, so that B_{max} does not occur at pressures greater than the saturation pressure.

Using this assumption, we can now calculate the value of P_i. Rearranging Eqn.9:14, we have

$$P_i = \frac{L(v/J) - P_e}{(v/J) + 1} \qquad (9{:}15)$$

Using Eqn. 8:22 to calculate v/J, we can determine P_i. Almost identical values are obtained if we use experimental v's instead of calculated v's from Eqn. 8:22. Both of these values of P_i are not much different from the values of P_i calculated from surface tension. The question remains, which values are preferable. A comparison of these values of P_i may be made by the use of the surface tension equation. Rewriting Eqn. 7:10, we have

$$\frac{\gamma}{a'} = 1 - \frac{b}{a'} T$$

or

$$\frac{b}{a'} = \frac{1 - \frac{\gamma}{a'}}{T} \qquad (9{:}16)$$

γ/a' can be obtained from Eqn. 8:18. Since we know the value of both P_i and L in the region where the surface tension is a straight line, the value of b/a' should be constant. Table 9.6 gives values of v, P_i and b/a' for the various options.

If we examine Table 9.6, we see that the values of b/a' from least-squares surface tension are much better than the others, but this is to be expected since we forced the values of γ to fit a straight line. Hence we reject these values of P_i, since they give negative P_e(max). Of the other two sets of results {those derived under the assumption that P_e(max) = P_e(sat)}, the standard error of the values derived using the theoretical v's is slightly smaller than that derived from the experimental value of v, but there is little real basis for choice. As can be seen, the calculation is extremely sensitive to minute changes. We shall use the theoretical values in the further calculations.

We can now calculate the values of $B_{max} = B_{sat}$ and $Z_{min} = Z_{sat}$ and also Z_{vap}, all at the saturation point. These values are shown in Table 9.7. As can be seen, the values thus obtained offer no surprises and normal helium 4 does not seem to be dissociated in the vapor, as shown by the value of Z_{vap}. Nevertheless, there exists a problem, since in evaluating this latter quantity the value of the covolume, B, is ignored.

172

Table 9.6

Normal helium 4

Values of P_i and b/a' for various options

	Experimental			Theoretical			From surface tension	
T°K	v cc/FW	P_i atm	$b/a' \times 10^2$	v cc/FW	P_i atm	$b/a' \times 10^2$	P_i atm	$b/a' \times 10^2$ *
2.5	27.61740	7.067143	4.49780	27.790580	7.072166	4.47286	7.8904	0.3621
3.0	28.35908	5.446426	3.67266	28.482145	5.449032	3.65847	6.0548	0.3595
3.5	29.49177	3.532046	3.21924	29.584445	3.533310	3.21017	3.9305	0.3592
4.0	31.13410	1.701617	3.40952	31.235181	1.702403	3.39955	1.9426	0.3518
4.2	31.97476	1.107592	3.97700	32.041291	1.108009	3.96053	1.3090	0.3707

average b/a'(exp) = 0.037552 ± 0.00504 (standard error)

average b/a'(theor) = 0.037421 ± 0.00498 (standard error)

average b/a'(from surface tension) = 0.003707 ± 0.000068 (standard error)

* These values are an order of magnitude less than the others. This is a result of the extreme sensitivity of the calculation.

Table 9.7

Normal helium 4

Values of B_{liq}, Z_{liq}, and Z_{vap} at the saturation point

T°K	P_i atm	P_e(sat) atm	$A \times 10^4$	B_{liq}	Z_{liq}	Z_{vap}
2.5	7.07217	0.10196	0.34974	0.101412	9.8603	1.036
3.0	5.44903	0.23957	0.22191	0.069517	14.3849	1.148
3.5	3.53331	0.46822	0.13079	0.041439	24.1318	1.208
4.0	1.70240	0.81123	0.08115	0.023041	43.4003	1.393
4.2	1.10801	0.98596	0.07567	0.018637	53.6554	1.480

$B_{max} = B_{sat}$, $Z_{min} = Z_{sat}$

This dilemma will probably be resolved in the future, when B is more precisely determined, since $(v-B) = v$ is a poor approximation.

THE SOLID STATE

The chief difference between liquids and solids lies in the fact that, while the structure of liquids is distinguished by the existence of inexact 5-symmetry, the structure of solids revolves around exact 6-(or 4-)symmetry. Because of this difference, the properties of solids are characteristically different from those of liquids. However, both solids and liquids are condensed phases and they seem to have certain properties in common.

THE TAIT-TAMMANN LAW AND SOLIDS

One of these common properties seems to be their conformity to the Tait-Tammann Law. Data suitable for testing the application of this law are not very abundant and much of the data that exist are flawed by inadequacies, uncertainties and inaccuracies. Nonetheless, some of the data are suitable. A good example of useful data are that of Bridgman on the alkali metals. Bridgman carried out and reported several series of runs on these substances. The runs all differ in the values that they contain, and Bridgman himself was uncertain as to which set should be considered most reliable. A partial analysis of the data of Bridgman on the alkali metals has been given by Ginell and Quigley. We believe that the criteria for good data are: (1) that a first approximation to the Tait-Tammann coefficients containing a positive value of J can be obtained, (2) that a convergent Deming adjustment can be carried out that gives values reasonably close to the values of the first approximation, and (3) that the theoretical volume or density calculated from these adjusted Tait-Tammann coefficients agrees fairly well with the experimental values.

The rationale behind these criteria is this: values of the coefficients to fit the Tait-Tammann equation can always be derived, at least graphically. In the first instance, these values can be considered only empirical constants fitting the equation to the data. If one attempts to arrive at the values of the coeeficients analytically, the problem is not as simple. If the data are not a reasonable set, then it is extremely difficult to get a reasonable set of coefficients. By "reasonable" I mean: first, in drawing the graph of a set of data containing scattered points, the eye tends to disregard outlying points. Analytically, these same points play a very important role unless specifically excluded. For this reason a bad set of data will often give better values with a graphical solution than with an analytical one. The Lagrangian interpolation subroutine usually uses 5 or 7 data points. This results in smoothing limited regions and sometimes introduces small systematic errors. If the data are reasonable, all these imprecisions have only a small effect on the resulting coefficients. Hence by "reasonable" I mean that the data should give only small imprecisions and result in a set of coefficients that conform to those of related sets of data or follow a pattern. Only after many determinations does the pattern emerge. In general, we found data with negative J's to be poor. Later determinations with better data on the same substance usually give us coefficients with positive J's. On the second criterion concerning the Deming adjustment, we have found that when the Deming adjustment does not converge to values close to the initial approximation, the coefficients are useless for reproducing the data and will give poor values of the derived quantities. The final criterion is that the coefficients reproduce the data with precision. This is the ultimate test. It is a necessary test but may not be a sufficient one, since good densities may be obtained without fulfilling the other criteria. Unless these criteria are fulfilled, we believe the coefficients to be without theoretical significance, although they may be useful in representing the data. As an example of the difficulty in choosing acceptable values see Table 9.8, which gives the results of an analysis of the various sets of Bridgman data on the alkali metals. As can be seen, Br.I gives high and inconsistant values of L, while Br.III gives, for no apparent reason, negative values for L in potassium. The values of Br.II seem to be the best set and after the Deming analysis show more consistancy. The values of L decrease with an increase in the atomic weight, as do the values of H. The values of J are not as consistent but also increase with the atomic weight when given in terms of cm^3/FW (see Table 9.9). Pending better reasons, we believe that these criteria are necessary to get consistent values of the coefficients.

THE ALKALI METALS

These criteria have been used in calculations on the data of the alkali metals. We use the determinations of Bridgman II, because we deem them most reliable. The reliability of data is very difficult to determine if no acceptable theoretical values are known; Bridgman's various sets suffer from this defect. Bridgman himself admitted in one of his papers that he was uncertain which of his data sets most accurately represented the facts, and consequently introduced various arbitrary adjustments into his compression values. The results of our analysis are listed in Table 9.9. In this particular determination (Br.II) values for cesium above the transition point of 23300 kg/cm^2 are also included. The analysis of these data yields values which, while converging, do not agree with the first approximation and hence are not given in Table 9.9.

The densities calculated from the Bridgman compressions and from the Tait-Tammann coefficients listed in Table 9.9 are given in Table 9.10. As can be seen from the standard error, the agreement in densities is excellent. On the other hand, as an example of the poor values for cesium above the transition pressure, this substance gives a standard error in the density of 1.306.

Table 9.8

Comparison of Tait-Tammann coefficients derived from first approximations of various sets of Bridgman data and the final value chosen after the Deming analysis, all at room temperature.

Metal	Data	J cm³/gm	L atm	$H \times 10^{-6}$ atm	comment
Lithium	Br.I	0.13135	152092.0		
	Br.II	0.38805	23769.0	2.9710	
	Br.III	0.40657	33561.6	2.9459	
	Br.II+III	0.37210	24211.7	3.6224	
	Br.II+Deming	0.39949	24875.38	2.7030	Acc.val.*
Sodium	Br.I	0.25933	24612.2		
	Br.II	0.18222	10313.8	3.0521	
	Br.III	0.21390	28094.1	2.3664	
	Br.II+III	0.18550	14416.8	3.2259	
	Br.II+Deming	0.20063	12815.72	2.1761	Acc.val.*
Potassium	Br.I	0.32762	21792.7		
	Br.II	0.18413	3455.65	2.1294	
	Br.III	0.13680	-6828.0	6.2631	
	Br.II+III	0.14521	-2911.1	5.1668	
	Br.II+Deming	0.20533	5410.64	1.4633	Acc.val.*
Rubidium	Br.I	0.44719	73493.8		
	Br.II	0.10566	2912.5	1.8065	
	Br.III	0.10839	6626.7	1.8434	
	Br.II+III	0.10446	3660.75	1.9800	
	Br.II+Deming	0.11961	5115.15	1.2107	Acc.val.*

* Acc.val. = Accepted value

THE INTERNAL PRESSURE OF SOLIDS

While a determination of the Tait-Tammann coefficients is a necessary first step in the elucidation of the structure of solids and the criteria we have chosen seem adequate to insure that coefficients have theoretical meaning, we would like to have some verification with independently derived data in order to create some confidence in these values. For this purpose we need the internal pressure of the solids. In the case of liquids we derived the internal pressure from the Tait-Tammann coefficients using the values of the surface tension (except for helium). The surface tension of solids, although it presumably exists, is not available by the methods used to derive the surface tension of liquids, hence this line of attack is closed to us. The approach we use to solve this problem is interesting.

If we examine Table 9.5 we notice that at the maximum in B, its value lies either a little above or somewhat below 1. To calculate B_{max} we need, however, the value of P_i, which we do not have. Let us then assume values of P_i and calculate $P_e(max)$ and B_{max} for a number of values of P_i. A set of such values is given in Table 9.11.

Looking at the values in Table 9.11 we see that well before the value of $B_{max} = 1$ is reached the value of P_e becomes negative. Since negative pressures are hypothetical, what is the meaning of these negative values of P_e? To understand this we must look back to the meaning as discussed for liquids. In liquids at $P_e(max)$ the breaking in bonds due to the distortion of the liquid's 5-symmetry structure was equal to the making of new bonds due to the rearrangement. This step occurred approximately where the system decreased in degree of association until B reached about 1. In solids, curiously, when B reaches the neighborhood of 1, but before it reaches it, P_e turns negative. Apparently, the mechanism in solids is not the same as that in liquids. This is to be expected. Solids are in exact 6-symmetry and there are very, very few voids in the structure that can be squeezed out by the application of pressure. Hence we can assume as a first approximation, at least, that the maximum in B for solids occurs at zero pressure. In other words, in solids, when pressure is applied no bonds (or very few) are broken but only new bonds are

Table 9.9

Tait-Tammann coefficients of the alkali metals* at room temperature

Metal	J cc/FW	L kg/cm^2	H kg/cm^2
Lithium	2.77285	24875.38	2702998.9
Sodium	4.61251	12815.72	2176073.5
Potassium	8.02877	5410.74	1463302.9
Rubidium	10.22306	5115.15	1210653.5
Cesium†	13.79424	4200.62	705655.84

* Data from Bridgman 1948 40,000 kg/cm^2 data
† Low pressure form, to transition at 23300 kg/cm^2

Table 9.10

Density as calculated from experimental values of the compression and as calculated from the Tait-Tammann coefficients at room temperature

Metal	P kg/cm^2	ρ_{exp}*	ρ_{calc}*	
Lithium	0	0.5340	0.5339	
	2500	0.5453	0.5451	
	5000	0.5556	0.5556	
	10000	0.5751	0.5754	
	15000	0.5937	0.5937	
	20000	0.6111	0.6108	
	25000	0.6271	0.6270	
	30000	0.6425	0.6423	
	35000	0.6570	0.6570	
	40000	0.6709	0.6712	Standard error = 0.000197 = σ
Sodium	0	0.9712	0.9707	
	2500	1.0048	1.0056	
	5000	1.0358	1.0373	
	10000	1.0931	1.0936	
	15000	1.1441	1.1433	
	20000	1.1896	1.1883	
	25000	1.2311	1.2299	
	30000	1.2695	1.2688	
	35000	1.3052	1.3055	
	40000	1.3377	1.3404	Standard error = 0.001321 = σ

Table 9.10 (continued)

Metal	P kg/cm^2	ρ_{exp}*	ρ_{calc}*	
Potassium	0	0.8700	0.8697	
	2500	0.9332	0.9330	
	5000	0.9833	0.9848	
	10000	1.0691	1.0696	
	15000	1.1408	1.1399	
	20000	1.2037	1.2016	
	25000	1.2596	1.2573	
	30000	1.3102	1.3087	
	35000	1.3560	1.3569	
	40000	1.3974	1.4024	Standard error = 0.00229 = σ
Rubidium	0	1.5300	1.5293	
	2500	1.6445	1.6494	
	5000	1.7434	1.7472	
	10000	1.9082	1.9073	
	15000	2.0416	2.0404	
	20000	2.1610	2.1572	
	25000	2.2680	2.2633	
	30000	2.3648	2.3615	
	35000	2.4519	2.4537	
	40000	2.5306	2.5414	Standard error = 0.00480 = σ
Cesium low pressure form	0	1.8730	1.8804	
	2500	2.0809	2.0689	
	5000	2.2258	2.2201	
	10000	2.4619	2.4668	
	15000	2.6685	2.6732	
	20000	2.8561	2.8567	
	23300	2.9806	2.9692	Standard error = 0.00688 = σ

* Rounded off to 4 decimal digits

Table 9.11

Values of B_{max} at P_e(max) for various assumed values of P_i. Sodium at room temperature.

P_i kg/cm^2	B_{max}	P_e(max) kg/cm^2
1250.0	6.09	41444.0
2500.0	5.60	36555.0
3750.0	5.09	31543.0
5000.0	4.56	26391.0
6250.0	3.99	21077.0
7500.0	3.40	15567.0
8750.0	2.76	9814.0
10000.0	2.07	3737.0
11250.0	1.29	-2821.0
12500.0	0.338	-10358.0
12812.5	0.0059	-12777.0

formed. This additional fact enables us to determine the internal pressure of solids to a first approximation.

Setting $P_e = 0$, Eqn.9:14 becomes

$$0 = (L - P_i)(v/J) - P_i$$

or

$$P_i = L/[(v/J)+1] \qquad (9{:}17)$$

Using P_i we can calculate A by Eqn.8:27, B_{max} by Eqn.9:4, and Z_{min} by Eqn.9:5. The results of this calculation are shown in Table 9.12.

Table 9.12

Values of P_i, A, B_{max} and Z_{min} for alkali metals at room temperature. Assumption P_e(max) = 0.

Metal	P_i kg/cm^2	A	B_{max}	Z_{min}
Lithium	20502.2	0.00161788	1.84125	0.54311
Sodium	10726.6	0.00096003	1.62733	0.61451
Potassium	4590.94	0.00056024	1.22899	0.81368
Rubidium	4324.15	0.00065336	1.46851	0.68096
Cesium (low pressure form)	3514.68	0.00097206	1.59408	0.62732

Looking at this table we see that the values of Z at $P_e=0$ are less than 1. This is strange, since the 1-mer in these calculations was taken as the atom. The only kind of dissociation possible is the breakup of the atom into a nucleus and an electron. We know metals have free electrons. Perhaps this is a measure of the extent to which this occurs.

ATOMIC RADII

Now that we have the value of A, we can calculate the value of B, the covolume or excluded volume of the particles, using Eqn.9:3. As can be seen from Table 9.13, the covolume B is almost equal to the volume, v, making the free volume, $(v-B)$,

Table 9.13

Values of the atomic radii of the alkali metals calculated from the covolume compared with the X-ray values obtained by Bragg and by Slater. Bridgman data at room temperature.

Metal	v cc/FW	B cc/FW	r Å	r Å (Slater)	r Å (Bragg)
Lithium	12.9998	12.8240	1.512	1.45	1.50
Sodium	23.6834	19.8228	1.748	1.80	1.77
Potassium	44.9617	38.1494	2.175	2.20	2.07
Rubidium	55.8865	47.2443	2.335	2.35	2.25
Cesium (low pressure form)	70.6802	59.1385	2.517	2.60	2.37

very small. One of the quantities we can now calculate is the atomic radii of the alkali metals.

The quantity that we must evaluate is the relationship of the covolume to the radii. We start with the knowledge that the alkali metals are an array of atoms in body-centered cubic symmetry (bcc). In the (bcc) array there are three particles along the diagonal of the unit cell that are touching each other (Fig.9.2). This makes the length of the diagonal $4r$, if r is the radius of a particle assumed to be spherical, and the distance is measured from the center of each particle. Since the length of the unit cell is l, using the Pythagorean theorem in three dimensions, we have

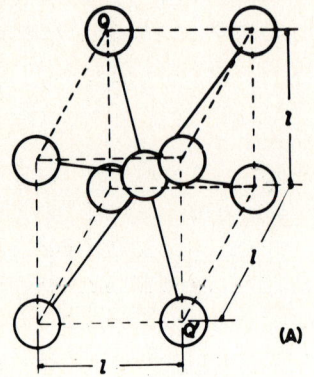 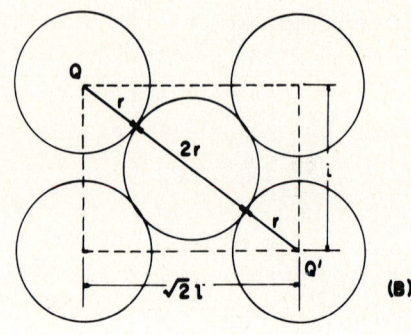

Fig.9.2. Body-centered cubic lattice (bcc).
    A - Perspective view - ball and rod model. Dotted lines are outlines of unit cell. Solid lines are line of contact. Balls are drawn small size to show spatial relationships. QQ' is the diagonal.
    B - The diagonal (110) plane. Balls are drawn full scale and touch. Dotted lines outline the plane. QQ' is the diagonal of the cube and is equal to $4r$. The side of the cube is l. Both views are in the same scale, except that balls are drawn smaller in view A.

$$(4r)^2 = 3l^2$$

from which

$$l = 4r/\sqrt{3}$$

Since in a (bcc) array there are two particles per unit cell, then the volume per particle or atom is

$$\text{volume per atom} = \frac{l^3}{2} = \frac{32r^3}{3\sqrt{3}}$$

and for an Avogadro number of particles it is

$$\text{volume per formula weight} = \frac{32r^3 N_0}{3\sqrt{3}}$$

This is the volume per formula weight of particle in an array of particles in (bcc) symmetry, and is equal to the value of B, which is the covolume or excluded volume of the substance per formula weight,

$$B = \frac{32r^3 N_0}{3\sqrt{3}}$$

Solving then for r gives

$$r = \left[\frac{3\sqrt{3}B}{32 N_0}\right]^{1/3} \qquad (9:18)$$

The corresponding quantity for a face-centered cubic array (fcc) is

$$r = \left[\frac{B}{N_0 4\sqrt{2}}\right]^{1/3} \qquad (9:19)$$

and for the simple cubic array (sc) is

$$r = (B/8N_0)^{1/3} \qquad (9:20)$$

Using Eqn. 9:18 we can calculate the radii of the alkali metals from the value of B. In Table 9.13 are given the results of such calculations, together with the values obtained by Slater and by Bragg from X-ray measurements. As can be seen from this table, the agreement is remarkable and justifies our confidence in these values of the Tait-Tammann coefficients and the theory in general.

OTHER SOLIDS

Data for other metals or solids that are suitable for this analysis are rather scant. Bridgman, in this same 40,000 kg/cm^2 set of data, gives values for germanium and cerium that seem suitable. The germanium data do not give a convergent Deming adjustment, although the values as determined by the slope and intercept (first approximation) give values for the density that fit the experimental data rather well (standard error = 0.00226). We regard this as just fitting of experimental data. We feel that the failure of the Deming adjustment to be calculable is an indication that the data are poor in quality, and hence we discard the results. With cerium, on the other hand, we get values of the coefficients that are all negative and of such magnitude that the densities we get are positive and fit rather well with a standard error of 0.00238. Again, from past experience, we reject these values as just values calculated to fit the curve.

Still, the low error in these density values and our ignorance about solids gives us pause and we shall consider these rejections tentative and await further calculations on other solids.

As an example of other data, we present calculations made on a set of values for sodium covering a wide range of temperatures. In Table 9.14 we give the values of J, L, and H. The calculation conforms to the criteria previously outlined.

Table 9.14

Tait-Tammann coefficients calculated for Beecroft and Swenson data.

T°K	J cc/FW	L atm	$H \times 10^{-6}$ atm	
Sodium				
20	4.075	12820	3.35816	
51	4.242	13462	2.84564	
77	4.219	13042	2.87028	
77	4.165	12844	3.02770	
115	4.188	12565	2.95442	
138.6	4.144	12087	3.08027	
171	4.251	12133	2.78102	
201	4.159	11188	2.98567	
204	4.408	12256	2.39134	
246	4.279	11146	2.67290	
297	4.478	11433	2.27229	
297	4.566	11796	2.11467	Standard error of all values of v
297	4.433	11314	2.36290	calculated from these coefficients
349	4.565	11123	2.11118	$\dfrac{\sum \sigma}{n} = 0.0102$
359	0.548	9768	2.90381	

Convergent Deming analysis are obtained and the v (theor) fit the experimental v. We compare these data to those of Bridgman by calculating the atomic radius, r, from these values of E. Table 9.15 we present such a comparison. As can be seen, the values agree quite well. The only value in Table 9.14 that seems suspect is the one at 359°K, which is an order of magnitude less in J.

Table 9.15

Comparison of Beecroft and Swenson data for sodium with those of Bridgman.

T°K	Investigator	v cc/FW	P_i atm	$A \times 10^{+4}$	B cc/FW	r Å
23	B&S 1st run	23.6994	9616.27	7.9969	19.933	1.752
23	B&S 2nd run	23.6950	9896.25	9.0135	19.866	1.750
23	B&S 3rd run	23.6814	9530.02	7.5505	19.947	1.752
Room	Br.II+Deming	23.6834	10381.64	9.6003	19.823	1.748

Many other substances could be of great theoretical interest but have not been investigated thoroughly. Quartz crystal gives an analytical first approximation by the Tait-Tammann analysis. The values of J, H, and L are all negative and the meaning of these results, especially theoretically, is not at all obvious. These data of Bridgman may just be poor. We have also investigated some of Bridgman's data on the glasses. They all give first approximations but only one gives a convergent Deming analysis. This is borax glass. The values for this substance are given in Table 9.16.

Table 9.16

Tait-Tammann coefficients for borax glass*. Bridgman data at room temperature.

J cc/FW	L kg/cm^2	H kg/cm^2
15.0604	17544.27	0.240130×10^8

* Assumed 1-mer - $Na_2B_4O_7$, F.W. = 201.22

The values of L are more closely comparable in magnitude to those of a solid than to those of a liquid. Again, as in solids, the value of P_e(max) becomes negative before the value of B_{max} reaches 1. Hence we have used the same method, Eqn. 9:17, to derive the value of P_i. P_i for borax glass is given in Table 9.17, together with some associated quantities. What the meaning of the value of approximately 8 for B_{max} or the value of ≈ 0.125 for Z_{min} means is by no means clear at this point.

Table 9.17

Values of P_i, A, and B_{max} for borax glass at room temperature. Assumption P_e (max) = 0.

P_i kg/cm^2	A	B_{max}	Z_{min}
15410.339	8.88659×10^{-5}	8.0109	0.124830

We are not even certain whether the data are correct; if they are not, then the values are purely accidental. Another set of values in these Bridgman data is for Pyrex glass. Here the J's are negative and according to our criteria we disregard this result. The values of J in the other glasses given in this set are all negative in the first approximation and no further work was done on them. Better values are needed. Other data for compression of glasses are given in the literature and perhaps some of them could be analyzed by this method, but this awaits further work.

IMPORTANT EQUATIONS

$$v = J \cdot \ln\left(\frac{H}{L + P_e}\right) \qquad (9:1)$$

$$B = \frac{(P_e + P_i) A v e^{v/J}}{N_w RT} \qquad (9:4)$$

$$(v - B) = A v e^{v/J} \qquad (9:3)$$

$$Z = \frac{w/M_0}{B} \qquad (9:5)$$

$$P_e = (L - P_i)\left(\frac{v}{J}\right) - P_i \text{ (at maximum)} \qquad (9:14)$$

$$r = \left[\frac{3\sqrt{3}B}{32N_0}\right]^{1/3} \quad \text{for (bcc)} \qquad (9:18)$$

$$r = \left[\frac{B}{N_0 4\sqrt{2}}\right]^{1/3} \quad \text{for (fcc)} \qquad (9:19)$$

$$r = \left[\frac{B}{8N_0}\right]^{1/3} \quad \text{for (sc)} \qquad (9:20)$$

REFERENCES

Allen, J.F. and Misener, A.D., *The surface tension of liquid helium*, Proc. Cambridge Philos. Soc., 34 (1938) 299-300.

Amagat, E.H., *Mémoires sur l'élasticité et la dilation des fluides jusqu'aux très hautes pressions*, Ann. Chim. Phys., 29 (1893) 503-574.

Beecroft, R.I. and Swenson, C.A., *An experimental equation of state for sodium*, J. Phys. Chem. Solids, 18 (1961) 329-344.

Bragg, W.L., *Arrangement of atoms in crystals*, Philos. Mag., 40 (1920) 169-189.

Bridgman, P.W., *Pressure-volume relationship for 17 elements to 100,000 kg/cm^2* (Br.I in text), Proc. Am. Acad. Arts Sci., 74 (1942) 425-5000.

Rough compressions of 177 substances to 40,000 kg/cm^2 (Br.II in text), Proc. Am. Acad. Arts Sci., 76 (1948) 71-87.

The compression of 39 substances to 100,000 kg/cm^2 (Br.III in text), Proc. Am. Acad. Arts Sci., 76 (1948) 55-70.

Brown, S.D. and Ginell, R., *The partial structure of glass and the reconstructive nature of devitrification processes*, Symp. on Nucleation and Crystallization in Glasses and Melts, (Eds. M.K. Reser, G. Smith, and H. Insley) Am. Ceram. Soc., 1962, pp.109-118.

Deming, W.E., *Statistical Adjustment of Data*, Dover, New York, 1965.

Ginell, R. and Quigley, T.J., *Compressibility of solids and Tait's Law I: P-V relationships in the alkali metals*, J. Phys. Chem. Solids, 26 (1965) 1157-1169.

Compressibility of solids and Tait's Law II: Atomic radii of the alkali metals, J. Phys. Chem. Solids, 27 (1966) 1173-1181.

Keller, W., *Helium 3 and Helium 4*, Plenum Press, New York, 1969.

Kennedy, G.C., Knight, W.L. and Holser, W.T., *Proprties of water, Part III: Specific volume of liquid water to 100 C and 1400 Bars*, Am. J. Sci., 256 (1958) 590-595.

Holser, W.T. and Kennedy, G.C., *Properties of water, Part IV: Pressure-Volume relationships of water in the range 100 C-400 C and 100-1400 Bars*, Am. J. Sci., 256 (1958) 744-753.

Kerr, E.C. and Taylor, R.D., *The molar volume and expansivity of liquid He^4*, Ann. Phys. (N.Y.) 26 (1964) 292-300.

Keesom, W.H. and Keesom, A.P., *Thermodynamic diagrams of liquid helium*, Physica, 1 (1934) 128-133.

Slater, W.L., *Atomic radii in crystals*, J. Chem. Phys., 41 (1964) 3199-3204.

Timmermans, J., *Physico-Chemical Constants of Pure Organic Compounds*, Elsevier Publishing Co., Amsterdam, Vol.1, 1950, Vol.2, 1965.

van Ure, A.Th., Keesom, W.H. and Kamerlingh Onnes, H., *Measurement of the surface tension of liquid helium*, Proc. K. Akad. Wet., 28 (1925) 956-962.

Handbook of Chem. and Physics, 51st edn., (Ed. R.C. Weast) 1970-71, Chem. Rubber Co., Cleveland, Ohio.

CHAPTER 10

THE CRITICAL STATE

One of the peculiar properties of the liquid state is that it ends in a critical point, the region beyond the critical point being called the fluid state. It seems logical that a theory that purports to explain the liquid state should also be able to explain the nature of the critical point, as well as the region surrounding it. The phenomenon of the critical point is unique to liquids. Although there has been some speculation as to whether solids possess a comparable point, none has ever been observed. Other phenomena in science have also been labeled critical phenomena, such as those in the field of ferromagnetism or binary solutions. Whether they are related in nature or by the fact that they occur at a more or less unique point is entirely speculative, since, while phenomenological aspects of these events are fairly well characterized, the reasons why they occur are not at all obvious*. In this section we will confine ourselves to a consideration of the critical point as it occurs in liquids.

THE PHENOMENOLOGICAL POINT OF VIEW

To describe the critical point briefly and succinctly one might say simply that below the temperature called the critical temperature, T_c, liquids can exist in equilibrium with vapors of the same substances, giving a heterogeneous system, while above this temperature only the homogeneous fluid state exists. Although this definition seems clear and unambiguous, it is not very operational, since it does not tell us how to distinguish the states. Clearly, the key seems to lie in the word heterogeneous; hence the fact that two distinct phases, the liquid and the gaseous, must coexist (this coexistence is not limited

---

* In this connection an interesting article by Sengers and Sengers should be consulted.

to the critical point). If one had a thin-walled tube containing a single phase, one would be hard pressed to tell whether it contained liquid or gas or even solid. If there were two phases present, there would be a boundary, although in fact the boundary might not be observable visually*. Our observational powers depend on the fact that two distinct phases of matter usually possess different indices of refraction. This difference is what makes the boundary observable. Liquids possess boundaries because they have fixed volumes, while gases do not. We can explore this fact with a mental experiment. Imagine we have a tube with a piston in a constant temperature bath confining a single-phase substance. We now proceed to move the piston so as to increase the volume of the confined space. If the single phase were a gas, we could move the piston out for the full length of the tube and still have a single phase. On the other hand, if the single phase were a liquid, after a certain amount of movement the piston would separate from the liquid and a space containing vapor would appear. We would now have a heterogeneous mixture of two phases. Assuming gravitational force is downward, the liquid phase would be at the bottom of the tube and the gaseous phase at the top. The pressure at this point would be the saturation pressure. If one now continued to move the piston and increased the volume, the pressure would remain fixed until all the liquid had been converted into gas, when the pressure would start decreasing with further volume increase.

An increase in temperature has the same effect as a decrease in pressure. Increasing the temperature of the liquid in the tube causes the piston to move outward so as to maintain a constant pressure. This expansion continues with increasing temperature until at one point the piston separates from the liquid and vapor forms. We are at the boiling point and further addition of heat does not cause the temperature to rise, but rather causes more and more of the liquid to be converted to gas till all the liquid changes state. The system, now homogeneous, rises

---

* A piece of glass placed in a particular liquid may disappear from view although the two phases remain present. We could prove that there are two phases, using a probe, even though our observational powers have failed us.

in temperature with further addition of heat. The results of many such experiments are summarized in Figs.10.1 and 10.2.

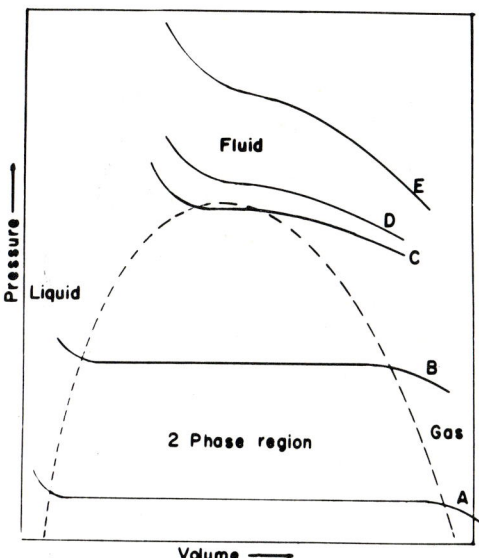

Fig.10.1. Schematic drawing of the relationship between pressure and volume for a pure substance. Isotherms A and B are well below the critical temperature. C is just below the critical temperature. D is just above the critical temperature. E is well above the critical temperature. For a substance like CO_2 the difference in temperature between an isotherm like A and one like E is approximately 1°C.

A simple method that can be used experimentally to demonstrate the existence of the critical point is to observe the liquid at its saturation pressure as the temperature is increased. In practice, this can be accomplished by placing a liquid in a tube, then evacuating and sealing it. The tube should only be partially filled and the liquid carefully chosen*.

---

* Before actually doing this experiment, the experimenter should carefully check the literature for exact directions, since the experiment is hazardous and has given rise to many accidents. The conditions and materials used must be very carefully chosen in the interest of safety. A bare outline is given here.

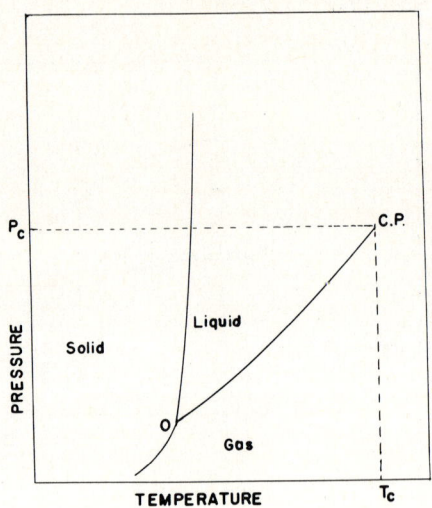

Fig.10.2. Schematic phase diagram of a pure substance. Projection on P-T plane. O is the triple point. $C.P.$ is the critical point. The liquid-gas equilibrium line ends at $C.P.$ In the diagram the liquid-solid equilibrium line has a positive slope, while the same line for water would have a negative slope.

One then places the tube in a heating bath and slowly raises the temperature, at the same time carefully observing the meniscus. As the temperature increases, the volume of the liquid also increases as it expands. The result of this expansion is an increase in the pressure in the tube. As the critical point is approached, the meniscus becomes progressively more indistinct and finally at one point disappears entirely. This point is commonly taken to be the critical point.

As can be imagined, the phenomenon is difficult to observe and several attempts have been made to try and pinpoint the exact temperature at which it occurs. One of these methods leads to quite puzzling results. It is well known that the density of the liquid in such a system approaches the density of the vapor as the critical point is approached. If one plots on the same graph the average of the liquid and vapor densities, one can see

that this line is straight (see Fig.10.3). The point where the
curve of the density is intersected by the curve of the average
is usually taken to be the value of the critical density. This
construction is known as the Law of Rectilinear Densities of
Cailletet and Mathias. Of course, the densities at the critical
point have not been observed, although measurements have been
made quite close to that point. Nonetheless, a certain amount
of extrapolation has always been done to join the gas and liquid
density curves, and a hypothetical average curve has been drawn.

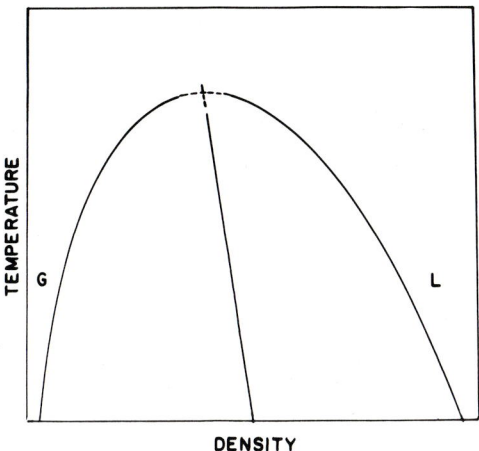

Fig.10.3. Law of Rectilinear Diameters. Section G is the gas region of the curve. L is the liquid portion. The inclined straight line is the average density. The intersection of the straight line and the curve is the critical point.

The experimenters, in trying to characterize the critical temperature, decided to try to use these observations. This was the experiment. They produced a large quantity of very small hollow glass bulblets, which had varying densities depending on the ratio of glass to air they contained. When a small quantity of these bulblets were introduced into the tube in which the critical point determination was to be carried out, some sank to the base of the tube, while others floated at the meniscus.

The tube was then evacuated and sealed and placed in the heating bath. As the temperature of the tube was raised, the density of the liquid decreased, however the density of the bulblets did not decrease at the same rate. Hence as the temperature increased, some bulblets detached themselves from the meniscus and sank to the bottom of the tube, their density now being greater than the density of the liquid. The expectation of the experimenters was probably that at the critical point the remaining bulblets would remain suspended in the bulk of the liquid and the lighter ones even rise to the top of the tube; but, in the perverse way of experiments, these expectations were not fulfilled, since an entirely different phenomenon occurred. At the disappearance of the meniscus, which could be observed, the bulblets remained suspended at the place in the tube where the meniscus had previously been. Further, if one carefully tipped the tube, the line of bulblets outlining the previous position of the meniscus, tipped as if suspended on a denser layer of liquid. In fact, they were suspended on a heavier layer of liquid, since if one at this point shook the tube, all the bulblets fell to the bottom; they were now denser than the remaining fluid. If one allowed the tube with the suspended bulblets to remain quietly at constant temperature, they gradually fell to the bottom as the density throughout the tube equalized (Fig.10.4)

The conclusion from this experiment is at variance with the extrapolation performed in applying the Law of Rectilinear Densities. Apparently the extrapolation of the density curves to make them join is not entirely justified. Various explanations of this curious phenomenon have been advanced. A prominent one is that at this point the system is not in equilibrium. This is an odd statement and seems more an evasion than an explanation. The explanation of the density difference in Chapter 5, which is based on association theory, seems to be more realistic. The reason for bringing up these experiments at this juncture is to emphasize the fact that there is some ambiguity about the exact value of the critical density. Therefore the question arises, what are the criteria defining the critical point?

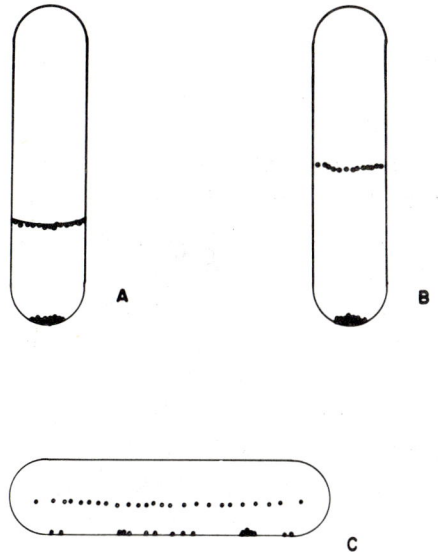

Fig.10.4. Critical point experiment.

    A – Below T_c. Beads floating at meniscus.

    B – At T_c. Meniscus disappears. Beads remain at former meniscus position.

    C – At T_c. Tube tipped. Level of beads tips showing lower layer to be more dense.

CRITERIA OF THE CRITICAL STATE

The semantics of the definition of the critical temperature seem clear. Below the critical temperature, liquids can exist in equilibrium with the gas phase; above the critical temperature liquids do not exist as a separate phase. The only trouble with this definition is that it is not operational and experimentally it does not tell us how to determine this point. In other words, it would be clearer if it would define how to determine when there is a liquid and if it gave criteria for determining when the critical state has been reached other than to say that there is a point where the liquid disappears.

THE DENSITY AND THE REFRACTIVE INDEX

The primary distinction betwee a liquid and a gas is that the liquid possesses a boundary, while the gas does not. What is the meaning of this boundary? A little speculative thought will show that by the word boundary we mean that, if we proceed through a single-phase substance and we come to a point where the properties change radically over a very short interval in the direction in which we are travelling, we have crossed a boundary. Because of this sharp alteration of properties, phenomena which depend on these properties are also altered sharply. Such a property is the refractive index. This property depends on the type of particles in its path and the density of such particles along the path. If there is a sharp alteration in these properties, then the refractive index changes suddenly, and this sudden change is what we see and recognize as a boundary. At the critical point, the meniscus, due to a sharp change in the refractive index, disappears after progressively becoming more indistinct, showing that there is no sharp change in the kind of particles in its path and there is no sharp change in the density. But there does exist a gradual and continuous change in these properties. This is why the bulblets remain suspended in the experiment outlined above. The density in the lower portion of the tube is greater owing to the fact that it started from a high density and is progressing to a lower density, while the upper section is less dense because it started from a lower density and proceeded to a higher one. The segregation is due to gravity and the only opposing force is diffusion, if we rule out external vibrations and minute temperature changes, which are really extraneous factors. These forces are not sufficient to insure that uniform composition be established. What then do we mean by an equilibrium state? Is thorough mixing and uniform composition a necessity for such a state, or is it sufficient that the state be invariant in time? In any case, what point should we take as the critical point? Most assuredly the density and refractive index are not positive determinators of this point.

THE SURFACE TENSION

The surface tension, although not usually considered a useful criterion of the attainment of the critical state, should be considered here, since there is a long history connecting it with the critical point. One of the reasons why it is not usually considered a criterion is the general experimental diffculty of making surface tension measurements in the neighborhood of the critical temperature. Originally, Eötvös postulated that the surface tension becomes equal to zero at the critical point. This concept seems intuitively correct since, when the liquid phase disappears, it seems logical to postulate that the force at the surface, the surface tension, should also disappear. Experimental observations by Eötvös in the critical region for several substances, when extrapolated to the then recognized critical points, seemed to confirm this postulation. Later work by Ramsey and Shields did not bear out this postulation. They found experimentally that the surface tension became zero below the critical point for six substances* and became zero above the critical point for three substances**. In both cases the final values of the surface tension were extrapolated. The six substances that fell below the critical point gave values that ranged from 5.9° to 8.9° below the critical temperature, averaging about 6°. Ramsey and Shields then, on this basis, modified Eötvös' equation to

$$\gamma (Mv)^{2/3} = k(T_c - 6 - T) \qquad (10:1)$$

It was more difficult for them to explain why the three other substances had extrapolated surface tensions above the critical temperature. Because of their concept of what the surface tension was and its connection with the word "surface", they could not imagine a surface tension could exist when the meniscus disappeared and there was apparently no surface. The only factor in their equation that was not fixed experimentally was M, the molecular weight of the substance. Hence, to reconcile these values to their equation, they ascribed to association the fact

---

* Ethyl ether, methyl formate, ethyl acetate, carbon tetrachloride, benzene, and chlorobenzene.

** Acetic acid, methanol, and ethanol.

that the surface tension became equal to zero above the critical temperature.

Going now to modern measurements, we find that at temperatures usually measured, the surface tension falls on a straight line (see Jasper for data). If these straight lines are extrapolated to γ = 0, we find that most of these extrapolated points fall below the critical point (see Table 10.1) although near it*. Whether the surface tension curves do in fact curve near the critical temperature is not experimentally certain. In any case, the values for diethyl ether and the first four alcohols do in fact extrapolate to a point above the critical temperature. Diethyl ether is one of the substances that, according to Ramsey and Shields, has a γ = 0 point below t_c. Although the association of these substances, as Ramsey and Shields claim, has historical weight, in light of our knowledge of the structure of liquids it seems dubious whether this is the explanation. From the standpoint of our discussion the surface tension is due to the disordered layer at the boundary between the gas and the liquid. The fact that the disordered layer encompasses the boundary and there is no longer a sharp demarcation point does not mean that the force no longer exists, although it may not be easily measurable by standard techniques. In any case, there seems to be an effect of structure on the magnitude of the depression below T_c to where the surface tension becomes equal to zero. This is apparent if we examine the values for the different $t_c - t(\gamma=0)$ and the ratio $T(\gamma=0)/T_c$. We shall examine a little later what this point γ = 0 means.

THE DIFFERENTIAL CRITERION

The differential criterion can be stated thus: at the critical point

$$(\partial P/\partial v)_T = 0$$
$$(\partial^2 P/\partial v^2) = 0 \qquad (10:3)$$

---

* One need not plot these values to get the extrapolated temperature, since the equation for the surface tension is
$$\gamma = a - bt = a' - bT$$
and at γ=0 the temperature is $t(\gamma = 0) = a/b$ or $T(\gamma = 0) = a'/b$.

The origin of this criterion lies in the van der Waals equation. If one writes the van der Waals equation

$$\left(P + \frac{a}{V^2}\right)(V - b) = RT$$

in expanded form in powers of V, one has

$$V^3 - V^2 \cdot (Pb - RT) + aV - ab = 0$$

This is a cubic equation in V and has three roots. If one plots the van der Waals equation on the same graph as the P-V plot of the substance, one sees that the first root corresponds approximately to the beginning of the condensation and the third root to the end of the condensation, while the second (middle) root is apparently devoid of meaning. At the critical point the three roots coalesce into an horizontal inflection point, where the conditions stated in Eqn.10:3 hold.

THE TAIT-TAMMANN EQUATION AND THE CRITICAL POINT

The concept stated in Eqn.10:3 is supported by the Tait-Tammann equation. If in the Tait-Tammann equation we set $(\partial P/\partial v)_T = 0$, one has

$$-(\partial P/\partial v)_T = (L + P_e)/J$$

at $\quad T = T_c, \quad \partial P_e/\partial v = 0$

Then $\quad P_c = -L_c$ $\hfill$ (10:4)

and $\quad \partial^2 P/\partial v^2 = 0$

We conclude that at the critical temperature the critical pressure is equal to $-L_c$. This conclusion is shown to be experimentally verifiable (see Table 10.2). Examination of Table 10.2 shows that the criterion of $\partial P_e/\partial v = 0$ will indeed lead to values of P_c that are closely comparable to the values of $(-L_c)$. More work on this point will undoubtedly confirm the validity of this equality. However, if we use L to determine the critical pressure and assume that at the critical point the mixture must be stirred, thereby taking the density as determined from the Law of Rectilinear diameters as the true value, we still have difficulties in this region of the critical point.

Table 10.1

Comparison of values of the temperature at which γ=0 and the critical temperature.

Substance	$t(\gamma=0)=a/b$ °C	t_c °C	$\dfrac{t(\gamma=0)}{t_c}$	$\dfrac{T(\gamma=0)}{T_c}$	$t_c - t(\gamma=0)$
Ethane	7.470	32.1	0.233	0.919	24.63
Propane	76.45	96.8	0.790	0.945	20.36
Butane	123.30	153.2	0.805	0.930	29.90
Pentane	165.59	197.2	0.840	0.933	31.61
Hexane	200.00	234.8	0.852	0.931	34.80
Heptane	225.51	267.0	0.845	0.923	41.50
Octane	247.35	296.2	0.835	0.914	48.85
Nonane	264.47	321.5	0.823	0.904	57.03
Decane	279.11	345.2	0.809	0.893	66.09
Undecane	293.67	369	0.798	0.883	75.33
Dodecane	306.68	387	0.792	0.878	80.32
Tridecane	318.04	410	0.776	0.865	91.96
Tetradecane	325.74	428	0.761	0.854	102.26
Pentadecane	336.02	444	0.757	0.849	107.98
Hexadecane	341.69	461	0.741	0.837	119.31
Heptadecane	349.88	476	0.735	0.832	126.12
Octadecane	355.72	491	0.724	0.823	135.28
Nonadecane	361.40	501	0.721	0.820	139.60
Acetic acid	297.59	321.6	0.925	0.960	24.01
Propanoic acid	288.82	337.6	0.856	0.920	48.78
Butyric acid	308.15	354.7	0.869	0.926	46.59
i-Butyric acid	292.17	336.3	0.869	0.928	44.08
Valeric acid	323.82	378.9	0.860	0.919	55.05
i-Valeric acid	307.90	360.7	0.854	0.917	52.78
Methanol	311.13	240	1.296	1.139	-71.13
Ethanol	289.06	243.1	1.189	1.089	-45.96
n-Propanol	337.97	263.7	1.281	1.138	-74.20
n-Butanol	302.57	287.0	1.054	1.028	-15.57
Ethyl acetate	226.44	250.1	0.905	0.955	23.66
Methyl formate	129.96	214.0	0.607	0.827	84.04
Carbon tetra-chloride	240.93	283.2	0.851	0.924	42.27

Table 10.1 (continued)

Substance	$t(\gamma=0)=a/b$ °C	t_c °C	$\dfrac{t(\gamma=0)}{t_c}$	$\dfrac{T(\gamma=0)}{T_c}$	$t_c-t(\gamma=0)$
Diethyl ether	208.37	193.6	1.076	1.032	-9.76
Chlorobenzene	302.02	359.2	0.841	0.910	57.18
Methyl ethyl ether	140.93	164.7	0.856	0.946	23.77
Stannic chloride	263.85	318.7	0.828	0.907	54.85
Hydrogen fluoride	132.33	188.0	0.704	0.879	55.67
Hydrogen bromide	63.01	80.9	0.701	0.926	17.89
Sulfur dioxide	136.44	157.4	0.867	0.951	20.96
Water	513.41	374.2	1.372	1.215	-139.21

Table 10.2

Comparison of $-L_c$ and P_c at the critical temperature.

Substance	$-L_c$ atm	P_c atm	Method
Helium I	2.57*	2.26	linear least squares on all points
Diethyl ether	36-48#	35.6	graphical
Water	220.8†	218.3	Lagrangian interpolation
Ammonia	111.40§	111.7	from empirical equation fitted to L values

* Calculated from data of Keesom and Keesom
\# The exact value is uncertain. It depends on how curve is drawn. Calculations from data of Amagat.
† Calculated from data of Holser and Kennedy.
§ Calculated from equation obtained by Kumagai, Date and Iwasaki.

THE SURFACE TENSION, L AND THE CRITICAL REGION

The value of L at the critical point is negative. Since it is positive at lower temperatures, somewhere between these two points it is zero. This zero value marks the beginning of what might be called the critical region. It is in this region that the surface tension becomes equal to zero. The question of interest is then, does this temperature where the surface tension becomes zero have any relationship to the temperature where L becomes equal to zero? These quantities are related through two basic equations. One of these is Eqn.8:18

$$P_i = \frac{\gamma}{a'} L \qquad (8:18)$$

From this equation we see that when L equals zero, the internal pressure, P_i, also equals zero, irrespective of the value of the surface tension, γ. Nevertheless, if γ equals zero at some other temperature, then there are two temperatures in this region where P_i is equal to zero. Such highly complex behavior seems unlikely to occur. Fortunately, we can investigate this easily, since both quantities, L and γ, are experimentally derived by totally independent methods.

THE AMMONIA DATA

Date has published considerable experimental work on ammonia. He has redetermined the values of the critical constants and published p-v data both in the critical region and in the adjoining liquid region, as well as computed the values of the Tait-Tammann coefficients. Using his experimental data, we have recomputed the values of the Tait-Tammann coefficients. The values we have obtained are given in Table 10.3. Certain of the series of data that Date gives do not give convergent Deming analyses and hence are omitted in the table. The ones that give convergent Deming analyses agree fairly well with his values, except the value at 100°C. This value agrees, however, with that determined by Tsiklis, who is cited by Date.

From the four values of L given in Table 10.3 we can make an attempt to determine the temperature at which $L = 0$. Plotting these values as in Fig.10.5, we find that because of the paucity of the data we can draw two curves with slightly different values of the temperature at which $L = 0$. Another method of finding this zero value is also available. Date gives an empirical quadratic equation fitting his values of L.

$$L = 8.2418 - 38.241 \, T + 0.043504 \, T^2$$

Setting $L = 0$ in this equation, we can solve for the T root of interest, getting 378.61°K or 105.36°C. This value lies between the two graphical values obtained above. We now have three possible temperatures to choose from at which $L = 0$.

Table 10.3

Values* of Tait-Tammann coefficients for ammonia (Date's data).

	Calculated from Date's data			Date values		Tsiklis values	
t°C	J cc/FW	L atm	H×10^{-4} atm	J cc/FW	L atm	J cc/FW	L atm
75	3.287	191.6	519.3	3.276	189		
100	3.531	32.32	357.3	3.442	26.2	3.654	33.6
125	3.969	−86.56	181.0	3.905	−87.6		
132.25	7.840	−111.36	0.1115	5.3704	−111.26		

t_c = 132.3°C
P_c = 111.7 atm
v_c = 72.05 cc/FW

* Data for other temperatures given by Date do not give convergent Deming analyses.

We find that the surface tension of ammonia has been determined at cryogenic temperatures from −75°C to −40°C. Since this cryogenic region is far from the critical region, we shall have to extrapolate, although such a procedure is far from being either safe or satisfactory. The compiler, Jasper, following the cue of the authors, Stairs and Sierko, fits their data with an empirical parabolic equation. Although from an empirical position this equation gives an excellent fit of the data, it is useless from a theoretical point of view, especially for extrapolation. Almost as good a fit to the data can be achieved by using a straight line. Better fits still can be obtained if the same cryogenic points are used and the value at the temperature where $L = 0$ is assumed to be the point where $\gamma = 0$. Since we have three choices as to the temperature where $L = 0$, we have three sets of values and correspondingly three linear regressions to carry out. The results are summarized in Table 10.4, where we give the original data from Jasper, the parabolic fit and the four sets of linear regressions: for the three values at the different temperatures where $L = 0$ and for the cryogenic points alone. We also give the value of r^2, which is the coefficient of the determination that indicates the quality of the fit. It can readily be seen that the value of $\gamma = 0$ at $L = 0$ at 105.9°C

Table 10.4

Values of γ (dynes /cm) vs. t°C for ammonia, using various least squares fits.

t°C	γ exp.	γ(0)	γ(1)	γ(2)	γ(3)	γ(4)
-75	43.39	43.39	43.56	43.83	43.78	43.79
-70	42.39	42.39	42.41	42.60	42.57	42.56
-65	41.34	41.34	41.27	41.38	41.36	41.35
-60	40.25	40.24	40.13	40.15	40.15	40.15
-55	39.10	39.10	38.98	38.93	38.94	38.94
-50	37.91	37.91	37.84	37.70	37.73	37.73
-45	36.67	36.67	36.69	36.48	36.52	36.52
-40	35.38	35.39	35.55	35.25	35.30	35.32
r^2*			0.99823	0.99974	0.99979	0.99981
a			26.3958	25.4565	25.6134	25.6570
b			0.22883	0.24494	0.24227	0.24150
a'			88.9016	92.3621	91.7898	91.6243

γ(0) = parabolic fit
γ = 23.41 - 0.3371t - 0.000943t^2 (equation given by Jasper)
γ(1) = cryogenic values only
γ(2) = cryogenic values + γ = 0 at 103.5°C
γ(3) = cryogenic values + γ = 0 at 105.36°C
γ(4) = cryogenic values + γ = 0 at 105.9°C

---

* r^2, the coefficient of determination indicates the quality of fit achieved by the regression. Values of r^2 close to 1.00 indicate a better fit than values close to zero.

$$r^2 = \frac{[n\sum x_i y_i - \sum x_i \sum y_i]^2}{[n\sum x_i^2 - (\sum x_i)^2][n\sum y_i^2 - (\sum y_i)^2]}$$

---

gives the best fit, since the r^2 value of this determination is closest to 1. It now seems that the surface tension is indeed equal to zero when the Tait-Tammann coefficient, L, equals zero. Since, however, one case does not prove the rule, we shall investigate several other substances for which data are available.

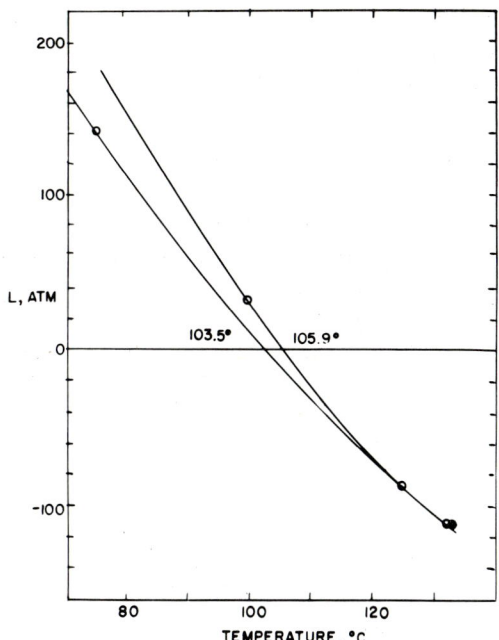

Fig.10.5. Ammonia. L vs. $t°C$ in the critical region. ● = t_c. $L = 0$ at 103.5°C or 105.9°C.

DIETHYL ETHER

From Table 10.1 we see that, if the straight line given by Jasper as the best fit to the surface tension data of diethyl ether is extrapolated to $\gamma = 0$, the temperature obtained is greater than the critical temperature. This is odd for several reasons, one of them being the fact that Ramsey and Shields cite this substance as being one at which the point $\gamma = 0$ lies approximately 6° below the critical temperature. If we examine the basis of Jasper's determination of the least squares coefficients, we find that it is based on only three values of the surface tension. This is hardly enough, especially when an extrapolation is to be carried out from points at 15, 25, and 30°C to the neighborhood of the critical temperature, which lies at 192.6°C.

We can find the temperature at which $L = 0$ graphically from the data in Table 8.2. This comes out to be 174.0°C. Assuming that at this temperature $\gamma = 0$, and using this point together with the values at 15, 25, and 30°C given by Jasper, we can redetermine the least squares coefficients. The fit thus obtained, which is shown in Table 10.5, is very good. Again, as with ammonia, it seems that $\gamma = 0$ when $L = 0$.

Table 10.5

Diethyl ether: surface tension vs. t°C, least squares determinations.

t°C	γ dynes/cm* (3 pts. only)	γ dynes/cm (3 pts. + γ=0 pt.)
15	17.56	17.72
25	16.65	16.61
30	16.20	16.06
174.0	--	0.016
r^2	0.999992	0.99976
a	18.92	19.3970
b	0.090314	0.11138
a'	43.5417	49.8218

---

* These values are those given by Jasper to fit his curve. They are used here as input.

WATER

Water is another compound that gives similar results, which, however, are not as clear-cut as those of ammonia and diethyl ether. We can obtain the temperature at which $L = 0$ from the Kennedy et al. data by interpolating graphically. The value 334.8°C that we thus derive is not quite certain, since it is difficult to fit the curve exactly. Nonetheless, we use this temperature as the point where $\gamma = 0$ and, using the experimental points of γ for water as given for the range 10°-100°C by Jasper, we can determine the least squares coefficients. Gittens has made a detailed study of the surface tension of water and reports the results of many authors. We report these details in Table 10.6, together with the calculated values of the surface tension at 25°C as determined by these coefficients. We also compare

Table 10.6

Values of the least squares coefficients for water and values of γ at various temperatures °C.

Author	a	$-b$	γ (25°)	γ (334.8°) $L = 0$	γ (374.2°) t_c	Range °C
Gittens (Drop vol.)	76.16	0.1569	72.24	23.63	17.45	5- 45
Gittens (Cap. rise)	76.13	0.1561	72.23	23.87	17.72	5- 45
Teitel'baum	75.78	0.1548	71.91	23.95	17.85	
Timmermans	75.23	0.1386	71.77	28.85	23.37	
Bordi-Vannel	77.01	0.1376	73.57	30.94	25.52	11- 44.5
Moser	75.78	0.1606	71.77	23.01	15.68	
Jasper	75.83	0.1477	72.14	26.38	20.56	10-100
This work*	79.58	0.2289	73.85	2.94	-6.07**	

$\gamma = a - bt$

* Using Jasper's values + $\gamma = 0$.

** If one follows Ramsey and Shields, this value agrees with their work, since it gives $\gamma = 0$ 6 degrees below the critical temperature.

---

these results to those obtained by using the value $\gamma = 0$ at $L = 0$ as an experimental point. It can be seen from the table that the value of γ at 334.8°C as derived from the least squares coefficients is not exactly zero. However, the value at the critical temperature now lies below t_c, instead of above it as given in Table 10.1, and in fact the zero value is about 6° below the critical temperature, in agreement with the conclusion of Ramsey and Shields. Considering all the inaccuracies in this calculation, it seems not inconsistent to say that for water $\gamma = 0$ when $L = 0$.

CONCLUSIONS

We thus have shown for three substances that $\gamma = 0$ at the same temperature where $L = 0$. While this is not definitive proof, it is at least an indication that the fact that the surface tension equals zero at $L = 0$ may be a general phenomenon. It remains to be determined what the values above $L = 0$ are.

From Eqn. 8:18 we have seen that at $L = 0$, $P_i = 0$. The other equation involving P_i is Eqn. 7:3

$$d = \frac{2\gamma}{P_i} \qquad (7:3)$$

Here d is the thickness of the disordered layer at the surface. Since, when $P_i = 0$, $\gamma = 0$, there is no problem with this equation, since now at this pint d is indeterminate. What happens when L becomes negative?

Since the equation for γ is $\gamma = a' - bT$ at temperatures above where $L = 0$ and $\gamma = 0$, γ is negative when bT is greater than a'. A negative γ and a negative L in Eqn. 8:18 give a positive P_i. This poses no problem except that it seems odd for P_i to go to zero and then increase again. However, a serious contradiction occurs when we examine Eqn. 7:3. A negative γ and a positive P_i give a negative d. Logically, this is impossible, since d is the thickness of the surface film and it is impossible to visualize what a negative d means. It is similarly difficult to conceive of what a negative γ means. Perhaps what happens is that γ goes to zero and there are no negative values of the surface tension despite the continuity of the surface tension equation. What about P_i then, when L becomes negative? At this point it is undefined.

There are some other peculiarities in this region. If we examine Eqn. 8:19

$$\frac{B}{v} = \frac{P_e + P_i}{P_e + L} \qquad (8:19)$$

we see that when P_i and L are both equal to zero $B/v = 1$. This means that, if this equation continues to hold above the temperature where $L = 0$, B becomes greater than v. This again is behavior that is difficult to conceptualize. The same conclusion can be reached in another way. From Eqn. 8:27

$$L - P_i = AH$$

we see that when L and P_i are zero, AH is zero. From the integrated Tait-Tammann equation H is not zero at this temperature, thus A must be equal to zero. Hence, from Eqn. 8:21,

$$1 - (E/v) = Ae^{v/J} \qquad (8:21)$$

$1 - (E/v) = 0$ or again that $E = v$.

To resolve these inconsistencies we must assume that between the temperature where $L = 0$ and the critical point the surface tension does not exist and that the equation derived for the lower temperatures in the liquid range does not apply. What is needed at this point is experimental data determined in the critical region between the temperature where $L = 0$ and the critical temperature. Such data will enable us to begin investigating this interesting region.

REFERENCES

Bordi, S. and Vannel, F., *Proprietà superficiale e variazioni strutturali dell'acqua*, Ann. Chim. (Rome), 52 (1962) 80-93.

Cailletet, L.P. and Mathias, E., *Densités des gaz liquéfiés et de leurs vapeurs saturées*, C. R. Acad. Sci.,102 (1886) 1202-1207.

J. Phys., 5 (1886) 549-564.

Date, K., *Studies on the P-V-T relationships of fluids at high pressures. I: The compressibility of ammonia*, Rev. Phys. Chem., Japan, 43 (1973) 1-15.

II: The P-V-T relations of ammonia in the neighborhood of the critical point and the critical values, Rev. Phys. Chem., Japan, 43 (1973) 16-23.

Eötvös, R., *Über den Zusammenhang der Oberflächenspannung der Flüssigkeiten mit ihren Molecularvolumen*, Ann. Physik, 21 (1886) 448-459.

Gittens, G.C., *Variation of surface tension of water with temperature*, J. Colloid Interface Sci., 30 (1969) 406-413.

Holser, W.T. and Kennedy, G.C., *Properties of water. Part IV: Pressure-volume-temperature relationships of water in the range 100-400°C and 100-1400 bars*, Am. J. Sci., 256 (1958) 744-753.

Jasper, J.J., *The surface tension of pure liquid compounds*, J. Phys. Chem. Ref. Data, 1 (1972) 841-1010.

Keesom, W.H. and Keesom, A.P., *Thermodynamic diagrams of liquid helium*, Physica, 1 (1934) 128-133.

Kumagai, A., Date, K. and Iwasaki, H., *Tait equation for liquid ammonia*, J. Chem. Eng. Data, 21 (1976) 226-227.

Moser, H., *The absolute value of the surface tension of water by the tearing-off method and its dependence on temperature*, Ann. Physik, 82 (1927) 993.

Ramsey, W. and Shields, J., *Über die Molekulargewichte der Flüssigkeiten*, Z. Phys. Chem., 17 (1893) 433-475.

The variation of molecular surface energy with temperature, Philos. Trans. Roy. Soc. London, 184A (1893) 647-675.

Sengers, J.V. and Sengers, A.L., *The critical region*, Chem. Eng. News, 46(24) (1968) 104-118.

Teitel'baum, B.Y., Gertolova, T.A. and Siderova, E.E., *Surface tension at various temperatures of aqueous solutions of lower alcohols*, Zh. Fiz. Khim., 25 (1951) 911-919.

Timmermans, J. and Bodson, H., *Surface tension of ordinary and heavy water*, C. R. Acad. Sci., 104 (1937) 1804-1807.

Tsiklis, D.S., *Compressibility of ammonia at pressures up to 10,000 atmospheres*, Dokl. Akad. Nauk. SSSR, 91 (1953) 889-890.

APPENDIX

LAGRANGIAN INTERPOLATION SUBROUTINE

This subroutine is written in Fortran IV for an IBM system. The data points presented to the machine must be in ascending order. The first part of the program is devoted to choosing the set of points to be used in the interpolation, the second part of the program to the interpolation proper. The program can easily be modified to use 9- or 11-point interpolation, but there is little use for such modification.

```
              SUBROUTINE INTERP (KI,MX,NNA,XIN,YIN,XOUT,YOUT)
C             KI=ORDER OF INTRPOLATION,3,5,OR 7
C             MX=NUMBER OF DATA POINTS
C             NNA=NO.POINTS OUTPUT
C             XIN=INDEPENDENT VARIABLE DATA INPUT
C             YIN=DEPENDENT VARIABLE DATA INPUT
C             XOUT=INDEPENDENT VARIABLE OUTPUT
C             YOUT=DEPENDENT VARIABLE OUTPUT
              DOUBLE PRECISION XIN(50),YIN(50),XOUT(50),YOUT(50),S(11),
             $G(11,11),SB(11),RR(50),TTA(50),XX,SA,SU,PR
              KJ=(KI-1)/2 +1
              DO 606 I=1,NNA
              IO=0
              MA=MX-KJ+1
              KZ=KJ-1
              DO 605 MN=KJ,MA
              IF(XIN(MN)-XOUT(I))604,608,660
     604      IF((MX-KZ)-MN)607,608,605
     605      CONTINUE
     606      CONTINUE
              GO TO 626
     607      WRITE(3,683)
     683      FORMAT('0 ERROR...NEG.QUANT.INTERP')
              STOP
     660      XX=(XIN(MN)+XIN(MN-1))/2.
              IF(XX-XOUT(I))608,608,661
     661      J=MN-1
              IQ=J
              IF(IQ-KJ)663,662,662
     663      J=KJ
              GO TO 662
     608      J=MN
     662      DO 639 JJ=1,MX
              IF(XIN(JJ)-XOUT(I))639,637,639
```

```
  637   YOUT(I)=YIN(JJ)
        IO=1
  639   CONTINUE
        IF(IO)607,609,605
  609   IO=1
        DO 630 IX=1,KI
        KY=J+IX-KJ
  630   S(IX)=XIN(KY)
        SA=XOUT(I)
        DO 610 LX=1,KI
        G(LX,LX)=SA-S(LX)
        LL=LX+1
        IF(LL-KI)613,613,614
  613   DO 611 L=LL,KI
        G(L,LX)=S(L)-S(LX)
  611   G(LX,L)=-G(L,LX)
  610   CONTINUE
  614   DO 615 IX=1,KI
        KY=J+IX-KJ
  615   SB(IX)=YIN(KY)
        DO 622 L=1,KI
        RR(L)=1.
        DO 621 K=1,KI
        IF(G(L,K))635,621,635
  635   RR(L)=RR(L)*G(L,K)
  621   CONTINUE
  622   TTA(L)=SB(L)/RR(L)
        SU=0.
        DO 624 K=1,KI
  624   SU=SU+TTA(K)
        PR=1.
        DO 625 K=1,KI
  625   PR=PR*G(K,K)
        YOUT(I)=PR*SU
        WRITE(3,681)I,J
  681   FORMAT('0 I=',I3,',J=',I3)
C       I=NO.POINTS IN DATA. J=CENTRAL POINT AROUND WHICH INTERP
        GO TO 605
  626   WRITE(3,682)KI
  682   FORMAT('0 END INTERPOLATION, ORDER=',I3)
        RETURN
        END
```

LAGRANGIAN DERIVATIVE SUBROUTINE

This program written in Fortran IV for an IBM system presupposes that the input data is equally spaced with an interval H. The maximum number of points is 50, although this can easily be changed. Since the first and last three points of the derivative have a larger error, these points can be skipped by using IND=1 in the call statement.

```
      SUBROUTINE DERIV (H,AV,NN,AT,AR,IND)
C     EQUALLY SPACED POINTS, INTERVAL=H
C     AV=DEPENDENT VARIABLE
C     NN=NO. OF POINTS, MAXIMUM 50
C     AT=DERIVATIVE D(AV)/D(INDEP)
C     AR=1/AT
C     ALLOWS SKIPPING 1ST AND LAST 3 POINTS. IND=1 FOR SKIP,
C     =2 FOR NO SKIP
      DOUBLE PRECISION AV(50),AT(50),AR(50),H,AA(49),AZ
      AA(1)=-1764.
      AA(2)=4320.
      AA(3)=-5400.
      AA(4)=4800.
      AA(5)=-2700.
      AA(6)=864.
      AA(7)=-120.
      AA(8)=-120.
      AA(9)=-924.
      AA(10)=1800.
      AA(11)=-1200.
      AA(12)=600.
      AA(13)=-180.
      AA(14)=24.
      AA(15)=24.
      AA(16)=-288.
      AA(17)=-420.
      AA(18)=960.
      AA(19)=-360.
      AA(20)=96.
      AA(21)=-12.
      AA(22)=-12.
      AA(23)=108.
      AA(24)=-540.
      AA(25)=0.
      DO 500 I=26,49
      J=50-I
  500 AA(I)=-AA(J)
      AZ=720.*H
      DO 540 I=1,NN
      AR(I)=0.
  540 AT(I)=0.
      GO TO (562,561),IND
  561 DO 545 N=1,3
```

```
            DO 544 I=1,7
            K=I+N*7-7
   544      AT(N)=AT(N)+AA(K)*AV(I)
   545      AT(N)=AT(N)/AZ
   562      K=NN-6
            DO 547 J=1,K
            L=J+3
            DO 548 I=1,7
            KK=I+21
            JJ=J-1+I
   548      AT(L)=AT(L)+AA(KK)*AV(JJ)
   547      AT(L)=AT(L)/AZ
            GO TO (564,563),IND
   563      N=NN-7
            NM=0
            DO 550 L=N,NN
            DO 551 I=1,7
            KK=I+28+NM*7
            JJ=NN-7+I
   551      AT(L)=AT(L)+AA(KK)*AV(JJ)
            AT(L)=AT(L)/AZ
   550      NM=NM+1
            IN=1
            INN=NN
            GO TO 565
   564      IN=4
            INN=NN-3
   565      DO 552 I=IN,INN
   552      AR(I)=1./AT(I)
            RETURN
            END
```

DEMING ANALYSIS SUBROUTINE

THE TAIT-TAMMANN COEFFICIENTS

One starts this calculation by using the Tait-Tammann equation in the form

$$-\left(\frac{\partial P}{\partial v}\right)_T = \frac{L}{J} + \frac{P}{J} \tag{A:1}$$

This linear equation is solved by least squares to get the best values of the slope, $1/J$, and the intercept, L/J, for the whole set of values of $\partial P/\partial v$ and P which has previously been derived. From these values of the slope and intercept the values of J and L are derived.

One now uses the integrated Tait-Tammann equation to derive the value of H for each set of values of P and v.

$$H = (P + L)e^{v/J} \tag{A:2}$$

From this set of values an average H is determined. Of course the error in these three values is not the same. The error in H is always much greater than that in J and L, because it is determined last. To even out the error in the three values we perform the Deming analysis. The input for the Deming analysis are the three Tait-Tammann coefficients J, L and H and the values P and v of the set.

The Deming analysis uses these values to obtain a better approximation of the values of J, L and H. This analysis should be repeated using the new values of J, L and H as input to get a second approximation. We repeat this process until the values of the coefficients converge. If the data is good this may take from 6 to 8 iterations. We generally use a tolerance of 10^{-6} to determine convergence, i.e., that the absolute difference between two successive values of each of the parameters divided by the value itself is less than 10^{-6}.

```
          SUBROUTINE DEMING (J,L,H,P,V)
C **   J IS IN GMS/CC, L & H IN UNITS OF DATA
C **   J,L,H ARE THE APPROXIMATION OF THE TAIT-TAMMANN COEFF,
C      P & V ARE THE PRES & VOL
          IMPLICIT REAL*8(A-H,O-Z), INTEGER*4(I-N)
          REAL*8 J,L
          DIMENDION P(50),V(50)
C
C      BEGIN ITERATION PROCEDURE
          WRITE(6,8100)
 8100  FORMAT('0 INITIAL APPROXIMATIONS FOR ITERATION='/)
          WRITE(6,8101)J,L,H
 8101  FORMAT('0 J=',D16.9,' L=',D16.9,' H=',D16.9,/)
  877     FLL=0.
          FLJ=0.
          FLH=0.
          FJJ=0.
          FJH=0.
          FHH=0.
          FOL=0.
          FOJ=0.
          FOH=0.
          DO 8500 I=1,N
          ZK=(P(I)+L)*DEXP(V(I)/J)
          FV=ZK/J
          FO=(ZK-H)/FV
          FJ=-V(I)/J
          FL=J/(P(I)+L)
C
C      NOTE EACH OF THE ABOVE HAS ALREADY BEEN DIVIDED BY FV,
C      NOTE ALSO FH=-1
          FLL=FLL+FL*FL
          FLJ=FLJ+FL*FJ
          FLH=FLH-FL/FV
          FJJ=FJJ+FJ*FJ
          FJH=FJH-FJ/FV
          FHH=FHH+1./(FV*FV)
          FOL=FOL+FO*FL
          FOJ=FOJ+FO*FJ
 8500  FOH=FOH-FO/FV
          FK=FLJ/FLL
          D4=FOJ-FK*FOL
          D3=FJH-FK*FLH
          D2=FJJ-FK*FLJ
          E3=FHH-D3*D3/D2-FLH*FLH/FLL
          E4=FOH-D3*D4/D2-FLH*FOL/FLL
          RESH=E4/E3
          RESJ=(D4-RESH*D3)/D2
          RESL=(FOL-RESH*FLH-RESJ*FLJ)/FLL
          H=H-RESH
          J=J-RESJ
          L=L-RESL
          WRITE(6,8101)J,L,H
          RETURN
          END
```

THE SUBROUTINE VAR

If one wishes to know the error in J, L and H after the application of the Deming subroutine, one uses the subroutine VAR. The call statement for this subroutine must be from the Deming subroutine. This program returns both the variance and the standard error of J, L and H.

```fortran
      SUBROUTINE VAR(FLL,FLJ,FLH,FJJ,FJH,FHH,P,V,N,L,J,H,VARL,
     $VARJ,VARH,SIGL,SIGJ,SIGH)
      IMPLICIT REAL*8(A-H,O-Z),INTEGER*4(I-N)
      REAL*8 J,L
C **
C **
C **   CALCULATES VARIANCE AND STANDARD ERRORS OF L,J & H IN
C **   EQUATION V=J*LOG(H/(P+L))
C **   REF. DEMING PG 167,SECTION62
C **
      DIMENSION P(50),V(50)
      COFAA=FJJ*FHH-FJH*FJH
      COFBB=FLL*FHH-FLH*FLH
      COFCC=FLL*FJJ-FLJ*FLJ
      DELTA=FLL*COFAA-FLJ*(FLJ*FHH-FLH*FJH)-FLH*(FLJ*FJH-FLH*
     $FJJ)
      SSS=0.
      DO 50 I=1,N
   50 SSS=SSS+(V(I)-J*DLOG(H/(P(I)+L)))**2
      FM=N
      CLOT=SSS/(DELTA*(FM-3.))
      VARL=DABS(COFAA*CLOT)
      VARJ=DABS(COFBB*CLOT)
      VARH=DABS(COFCC*CLOT)
      SIGL=DSQRT(VARL)
      SIGJ=DSQRT(VARJ)
      SIGH=DSQRT(VARH)
      RETURN
      END
```

CONVERGENCE

As stated the convergence of the values using the Deming analysis is quite rapid, six to eight iterations being sufficient to bring the difference to within 10^{-6}. The tolerance, however, should not be set smaller than this unless the data are certain within more than usual limits. Smaller tolerances result in the convergence depending on the round-off error of the original data. Occasionally the convergence is very slow. In such cases a method of improving convergence is useful. There are several methods in common use, their utility depending on the character of the slow convergence. When the nature of the slow convergence appears to be like a geometric progression, we have improved convergence by applying Eqn. A:3 to each variable independently.

$$a^* = a_K - \frac{(a_K - a_{K-1})^2}{(a_K + a_{K-2} - 2a_{K-1})} \qquad (A:3)$$

where a^* = value with improved convergence
a_K = last value of series
a_{K-1} = next to last value of series
a_{K-2} = second value before last in series

Following the application of the method of improving the convergence it is necessary to carry out another Deming analysis. The reason for this is that the improvement of convergence is done on each variable independently, the Deming analysis evens out the error.

Finally it is important to remember to carry out the calculations with sufficient digits, many in excess of the "significant" figures in the data. We have usually used double precision which on the IBM computer means that the calculations are carried out to 16 digits, although we do not report this many and only rarely print them. Use of a lesser number of digits results in a falsification of the results leading to erroneous conclusions. This is especially true in the interpolation and derivative subroutines. We once carried out a test of finding the derivative to a set of data using in separate calculations different numbers of digits. We first determined the derivative using 26 digits. Then we repeated the calculations using successively fewer and fewer digits. We looked at the answer to 10 digits. If we used less than 15 digits the answer to 10 digits changed. Hence our recommendation that double precision be used.

REFERENCES

Buckingham, R.A., *Numerical Methods*, Pitman, London, 1957.

Deming, W,E., *Statistical Adjustment of Data*, Dover, New York, 1964.

Nielsen, K.L., *Methods in Numerical Analysis*, Macmillan, New York, 1956.

Salzar, H.E., *Lagrangian Derivatives*, Appl. Math. Series 2, Natl.Bur. Stand. (USA), Washington, D.C., 1948.

Zaguskin, V.L., *Handbook of Numerical Methods for Solution of Equations*, Pergamon Press, Oxford, 1961.

AUTHOR INDEX

Alani, G.H., 87
Allen, G., 118, 120, 125, 126
Allen, J.F., 170, 187
Amagat, E.H., 135-139, 142, 147, 149, 153, 169, 187
Anderson, R.J., 127
Arons, A.T.S., 154

Beckett, C.W., 87
Beecroft, R.I., 184, 185, 187
Benedict, W.S., 87
Bernal, J.B., 65-67
Bethe, H., 154
Bird, R.B., 128, 154
Blatz, P.J., 28, 34
Bodson, H., 210
Boks, J.D.A., 14, 50, 126
Boltzman, L., 14
Bondi, S., 207, 209
Bragg, W.L., 181, 187
Bridgman, P.W., 173, 185, 189
Bromwich, T.J.I'A., 50
Brown, S.D., 128, 153, 187
Buckingham, R.A., 219

Cailletet, L.P., 193, 209
Carl, H., 128, 153
Clark, A.L., 95
Clausius, R., 7, 14
Clusius, K., 87
Coates, J.E., 87
Counsell, J.F., 127
Curtis, C.F., 128, 154

Date, K., 202, 209, 210
Davies, R.H., 87
Deming, W.E., 134, 140, 141, 153, 187, 219
Desai, 125, 126
Diesbergen, U., 87
Dorsey, N.E., 87
Drost-Hansen, W., 152, 153

Eötvös, R., 197, 209
Evans, W.H., 113
Eyring, H., 95

Fano, L., 87
Forsythe, W.E., 87

Gee, G., 118, 120, 126
Gertalova, T.A., 210

Gibson, R.E., 128, 153
Ginell, A.M., 95
Ginell, R., 14, 34, 50, 66, 95, 128, 154, 174, 187
Gittens, G.C., 207, 209
Glasstone, S., 14
Gleiser, M., 126
Guttman, L., 66

Halverson, R.R., 154
Hawkins, D.T., 126
Hayward, A.T.J., 128, 129, 154
Hildebrand, J.H., 117, 126
Hilsenrath, J., 87
Hirai, N., 95
Hirschfelder, J.O., 128, 154
Hoge, H.J., 87
Holser, W.T., 135, 138, 142, 147, 154, 187, 207
Hultgren, R., 125, 126
Hust, J.G., 87

Iwasaki, H., 210

Jaffe, I., 113
Jasper, J., 118-120, 126, 198, 203, 206, 209
Jeans, J.H., 14

Keesom, A.P., 141, 154, 188, 209
Keesom, W.H., 141, 154, 156, 170, 188, 209
Keller, W., 187
Kelley, K.K., 126
Kennard, O., 148, 154
Kennedy, G.C., 135, 140, 142, 147, 154, 158, 169, 187, 207
Kerr, E.C., 158, 187
Kirk, R.E., 87
Kirkwood, J.G., 144, 154
Kirsch, A.S., 95
Knight, W.L., 135, 140, 142, 147, 154, 187, 207
Knopp, K., 50
Kobe, K.A., 87
Kudchaker, A.P., 87
Kumagai, A., 210

Lange, N.A., 87
Lee, D.A., 127
Lees, E.B., 127
Levine, S., 113

Loeffler, O.H., 153
Lynn Jr., R.E., 87

Macdonald, J.R., 128, 154
Martin, J.F., 127
Masi, J.F., 87
Mathias, E., 193, 209
Meijering, J.L., 66
Miller, S.A., 87
Misener, A.D., 170, 187
Moser, H., 207, 210

Natanson, L., 14
Nielsen, K.L., 219
Nuttall, R.L., 87

Onnes, Kamerlingh, 9, 10, 11, 14, 16, 48, 50, 120, 170, 188
Orstrand, C.van, 46, 50
Othmer, D., 87

Quigley, T.J., 128, 154, 187

Ramsey, W., 197, 198, 207, 210
Ree, T., 95
Richardson, J.M., 144, 154
Richter, F., 87
Roder, H.M., 87
Rossini, F.D., 87, 113
Rusanov, A.J., 117, 127, 144, 155

Salzar, H.E., 219
Scott, R.L., 117, 126
Sengers, A.L., 189, 210
Sengers, J.V., 189, 210
Shields, J., 197, 198, 207, 210
Siderova, E.E., 210
Slater, W.L., 181, 188
Smith, C.S., 66
Stewart, R.B., 87
Swenson, C.A., 184, 185, 187

Tait, P.G., 128ff, 155ff
Tammann, G., 128ff, 155ff
Taylor, R.D., 158, 187
Teitel'baum, B.Y., 207, 210
Thomson, J.J., 14
Timmermans, J., 120, 124, 127, 158, 160, 163, 188, 207, 210
Tobolsky, A.V., 34
Touloukian, Y.S., 87
Tsiklis, D.S., 202, 210

Ure, A.Th.van, 170, 188

Vannel, F., 207, 209
Varde, E., 87

Wagman, D.D., 113, 126
Weigand, K., 87
Wilson, G.J., 118, 120, 126
Wohl, A., 128, 155

Wooley, H.W., 87

Zaguskin, V.L., 219
Zwolinski, B.J., 87

SUBJECT INDEX

A, the constant, 132, 144, 180
Acetic acid, 197, 200
Acetone, 137
Acetophenone, 118, 120
Aggregation, 71, 165
    equation, 19, 20
    j-mers, 15
    rate of, 74
    saturation point, 157
Alkali metals, 173ff
Allyl alcohol, 137
Alpha-mer, 76, 77-81, 87, 88, 148
Ammonia, 201-206
Argon, 125
Array, 3-hole, 71, 72
    4-hole, 72
Atomic radii, 181

B, the factor, 162, 166-168
Benzene, 197
Boiling point, 10
Bonds, 70, 71, 75, 90, 93, 106, 108
    broken, 106, 112, 167ff
    formation, 72, 74
    stability of, 76
Boundary, gas-liquid, 198
Boyle's Law, 1, 2, 4, 99
isoButanol, 120, 122, 124, 127

Carbon disulfide, 109, 138
Carbon tetrachloride, 197, 200
Cerium, 183
Cesium, 178-181
Chlorobenzene, 197, 201
Close packed forms, 54
Collision, 69, 71, 73-75, 100, 101
    distance, 39
    instantaneous, 39
    statistical nature, 38
Compressibility factor, 88
    isothermal, 133
Condensation point, 77
Constant A, the, 132, 144, 180
Constant H, the, 133, 134, 143
Constant J, the, 132, 133, 143
Constant L, the, 130, 133, 142, 143, 144, 199, 201
Convergence, 218
Corresponding states, 10, 110-111

Covolume, 38, 41, 42, 44, 83, 84, 103, 145-148, 160, 161
Critical point, 10, 81, 82, 85
    pressure, 199
    state, 80, 189ff
Crystallization, 88-91

Defect volume, 84
Degradation, 69
    equation, 19, 20, 21
    j-mers, 16
Degree of association, average, 31
Deming analysis, 134, 135, 141, 183, 202, 203, 215, 216
Density, 158-178, 196
Diethyl ether, 118, 120, 122, 124, 127, 135, 138, 142, 143, 149, 151, 158, 160, 163, 166, 169, 197, 198, 201, 205, 206
Differential coefficient, 198
Discontinuous case, 79
Dissociation, 161, 164

Energy, 97, 99
    association, 145
    containment, 102
    stress, 101, 102
Enthalpy, 122, 123
Entropy, 96ff
Equation of state, 114, 129
    association, 35ff, 103, 113, 115
    Dieterici, 88
    liquid, 114, 115
    reduced, 10
    Tait-Tammann, 156ff, 173
    virial, 9, 10, 44, 49, 115
    Waals, van der, 7, 88, 114, 115, 199
Equilibrium constant, 22-28, 32
Equipartition of energy, law, 6
Ethanol, 136, 142, 147, 158, 163, 169, 197, 198, 200
Ethyl acetate, 197, 200
Ethyl bromide, 137
Ethyl iodide, 137

Factor B, the, 151ff
First law of thermodynamics, 96-98
Forces, 54

covalent, 52
electromagnetic, 51
electrostatic, 51

Gay Lusac's Law, 1, 2, 3, 4
Germanium, 183
Glass, 185
    borax, 185, 186
    Pyrex, 186

H, the constant, 133, 134, 143
Halogens, 108
Helium, 105, 108, 112, 141, 159, 170, 201
Hexane, 118, 120
Hydrocarbons, 200
Hydrogen, 112, 168
Hydrogen halides, 108, 112, 201
Hydrogen nitrate, 109, 111
Hydrogen sulfate, 108, 110. 111

Ideal gas law, 2, 7, 38
    Kinetic derivation, 4
Internal pressure, 115, 118, 122, 125, 126, 143, 202
    of solids, 161, 177

J, the constant, 132, 133, 143

Kinetic derivation of gas law, 36
Kinetic energy, 6, 75, 101, 102
    Ideal gas, 103
Kinetic molecular theory, 1, 3, 11, 13, 99

L, the constant, 130, 133, 142-144, 199, 201
Lagrangian interpolation, 133, 211
    differentiation, 133, 213, 219
Linear forms, 54, 75
    structure, 112
Liquid state, 112ff, 156ff
Lithium, 176, 178, 180, 181

Melting, 92, 93
Mercury, 113
Methanol, 136, 147, 150, 155, 162, 168, 197, 198, 200
    Bond distance, 150
    molecular volume, 148ff
Methyl ethyl ether, 201
Methyl formate, 197, 200
Model building, 56, 61

Nitrogen, 105, 124
Nitrogen oxides, 109
Noble gases, 108, 111, 112
Notation, 16

N, 28
C, 32
Nucleation, 88-91
    heterogeneous, 90
    homogeneous, 90ff

Opalescence, 81
Oxygen, 108, 124

Pentane, 118, 120
Phase change, 68ff
    diagram, 192
Phosphorus trichloride, 137
Potassium, 176, 178-181
Pressure, internal, 115, 118, 122, 125, 126, 143, 202
    saturation, 156
n-Propanol, 118, 120, 122, 124, 127, 137, 147

Quartz, 185

Rate equations, 18
Rectilinear diameters, 193
Refractive index, 80, 190
Rubidium, 176, 178-181

Second law of thermodynamics, 98ff
Series, reversion of, 44
Silico fluorides, 108
Sodium, 176, 178, 180, 181
Solid state, 173ff
Stannic chloride, 201
Sublimation, 92
Sulfur dioxide, 108, 111, 201
Sulfur trioxide, 108, 111
Supersaturated vapor, 93, 94
Surface, area, 145
    energy, 117, 145, 147
    film, 144, 147
        thickness, 116-118, 208
    layer, 115-117
    tension, 114ff, 164, 197ff, 204, 206
    volume, 117
Symmetry, 63-66, 83, 89-93

Tait-Tammann coefficients, 136, 176, 178, 203, 204
    equation, 128, 129, 156ff, 173, 199
Temperature, concept, 1
    critical, 9, 189ff
    meaning, 43
    scales, absolute, 2
    scales, practical, 2
Transition, gas-liquid, 73
    gas-solid, 92

　　　　　liquid-gas, 78, 105, 108-110
　　　　　liquid-solid, 88
　　　　　point, 66
　　　　　solid-gas, 92
　　　　　solid-liquid, 106, 108-110
　　　　　solid state, 93ff
　　　　　state, 101
Trichlorethane, 124

Unimer, critical number, 112
　　　　　multiply bonded, 76, 81, 106, 116
　　　　　simply bonded, 76, 81

Value of Z, 162, 167, 168
Var, subroutine, 217
Virial coefficients, 49
　　　　　equation, 9, 44
　　　　　　　closed form, 51, 115
Volume, free, 38, 104
　　　　　saturated, 156ff

Water, 108, 134, 135, 139, 140, 142,
　　　　146, 147, 151, 152, 159, 162,
　　　　165, 168, 201, 206, 207

Z, the value of, 162, 167, 168

QD
503
G55

JUN 21 1979